BIOCHEMISTRY RESEARCH TRENDS

INOSITOL

SYNTHESIS, FUNCTIONS AND CLINICAL IMPLICATIONS

BIOCHEMISTRY RESEARCH TRENDS

Additional books in this series can be found on Nova's website under the Series tab.

Additional e-books in this series can be found on Nova's website under the e-book tab.

BIOCHEMISTRY RESEARCH TRENDS

INOSITOL

SYNTHESIS, FUNCTIONS AND CLINICAL IMPLICATIONS

HENRIQUE ROCHA
AND
MARINA CARDOSO
EDITORS

NOVA BIOMEDICAL

New York

For permission to use material from this book please contact us:
Telephone 631-231-7269; Fax 631-231-8175
Web Site: http://www.novapublishers.com

NOTICE TO THE READER

Additional color graphics may be available in the e-book version of this book.

Library of Congress Cataloging-in-Publication Data

ISBN: 978-1-62808-520-4

Library of Congress Control Number: 2013945020

Published by Nova Science Publishers, Inc. † New York

Contents

Preface

In this book, the authors present topical research in the study of the synthesis, functions and clinical implications of inositol. Topics discussed include hypoxia and angiogenesis related myo-inositol phosphates; inositol derivatives in mycobacterium tuberculosis; inositols and epilepsy; the contribution of inositol trisphosphate signaling to cardiac pathologies; the role of inositol 1,4,5-triphosphate receptor in hepatorenal syndrome; the pathophysiologic significance of inositol phosphates (IP) in Alzheimer's and bipolar disorder; design and synthesis of biotinylated inositol phosphates; the health benefits of inositol and inositol hexaphosphate; and the cellular functions of inositol.

Chapter 1 - One of the most fundamental physiological processes in the aerobic organisms is the delivery of oxygen to all tissues. Deprivation of oxygen leads to hypoxia which is a pathological condition playing a central role in numerous types of diseases, such as cancer (solid tumors), cardiovascular diseases and anemia. In this chapter the authors review the synthesis of a number of *myo*-inositol phosphates and their biological evaluation regarding hypoxia and angiogenesis in an independent or a combined relationship. Some of the compounds were found to be involved in the increased oxygen delivery by free hemoglobin or whole red blood cells, in the suppression of hypoxia-induced factor 1α (HIF-1α), in the inhibition of the phosphatidylinositol 3-kinase (PI3K) pathway, and in the reduction of angiogenesis. Consequently, the eradication or growth reduction of solid tumors, the prevention of sever anemia caused by sickle cell disease, and the enhanced exercise capacity of subjects with severe heart failure have been demonstrated. These results have evinced a leading role for *myo*-inositol phosphates in the treatment of diseases related to hypoxia and angiogenesis.

Chapter 2 - Inositol (Ins) is a polyol used for the preparation of Ins derivatives that carry out essential functions in eukaryotes. Although rarely found in bacterial species there are Ins derivatives that play critical roles in actinomycetes, a group of Gram-positive bacteria that includes *Mycobacterium* species such as *M. tuberculosis*, the causative agent of tuberculosis (TB). Consequently, there is much interest in understanding the functions and biosynthesis of Ins derivatives in mycobacteria to gain insights that can be used for the development of therapeutic agents. Mycobacteria can either obtain Ins from the environment through Ins transporters or synthesize it *de novo* via a two-step enzymatic pathway. Cellular Ins is converted enzymatically into phosphatidylinositol (PI), a precursor for more complex glycolipids, or incorporated as L-*myo*-inositol 1-phosphate (Ins-1P) into the small molecule mycothiol (MSH). These Ins derivatives carry out critical functions required for mycobacterial viability and virulence. The mycobacterial cell envelop contains an abundance of mannosyl-phosphatidylinositol containing glycolipids and lipoglycans, including phosphatidylinositol mannosides (PIMs) lipoarabinomannan, (LAM), and lipomannan (LM). Importantly, PIMs, LAM, and LM have been shown to function as immunomodulatory molecules, promote the entry of mycobacteria into phagocytic cells, and regulate phagosome maturation. Mycobacteria also produce the Ins derivative MSH, which is used by mycobacteria as the primary reducing agent and in the detoxification of xenobiotics. Due to these important functions, mycobacterial Ins derivatives are of therapeutic interest for vaccine development and as sources of potential targets for drug development. This chapter reviews mycobacterial uptake of Ins and biosynthesis of Ins derivatives, known functions of Ins derivatives in *M. tuberculosis*, and the progress to date on targeting Ins derivatives for vaccine and drug development.

Chapter 3 - Epilepsy is a heterogeneous group of disorders. It is the most common neurological disorder after stroke. Despite achieved progress in the treatment of epilepsy, about one third of the patients with epilepsy are resistant to existing pharmacotherapies. The current treatment of epilepsy focuses exclusively on preventing or suppressing seizures, which are the symptoms of the underlying disease.

The most important challenge is to prevent epileptogenesis, the process by which brain becomes epileptic. Drugs that prevent the development of epilepsy are not yet available.

The central nervous system is characterized with relatively high concentrations of myo-inositol (MI) as well as the means to synthesize it. MI serves not only as a precursor molecule for inositol lipid synthesis, but also as a physiologically important osmolyte. Alterations in MI deposition may playrole in a number ofneuropathological conditions, either as a physiologically important osmolyte or as a precursor molecule for phosphoinositide synthesis.

Several lines of evidence indicate the involvement of changes in inositols and other osmolytes in epilepsy and the effects of inositols in the regulation of induced seizures. The authors have revealed that the water extract of the medicinal plant *Aquilegia vulgaris* (a plant widely used in Oriental folk medicine as antiepileptic and soporific treatments) contains compounds altering binding of ligands to the benzodiazepine andgamma-aminobuturic acid (GABA) binding sites of the GABA-A receptors. The authors have identified two such compounds of this extract: (1) (MI) and (2) sleep-inducing lipid oleamide. Further the authors have shown that MI and scyllo-inositol (SCI) pretreatment significantly decreases the severity of seizures induced either by pentylentetrazolium or kainic acid (KA). As these effects were achieved by physiological concentrations of the administered inositols, the authors have hypothesized that MI and SCI could represent endogenous anti-convulsants.

Our original data showed that MI also could interfere with the process of epileptogenesis. The authors have first induced status epilepticus by KA and then applied MI daily treatment. The authors have shown that a 28-day MI treatmentsignificantly attenuates biochemical changes associated with the process of epileptogenesis in the hippocampus, and restores the amount of some drastically reduced proteins to the normal level. To these proteins belongs the GLUR1 subunit of glutamate receptors, calcium/calmodulin dependent protein kinase II (CaMKII), a $\gamma 2$ subunit of GABA-A receptors. The obtained results indicate that MI treatment could at least modify the epileptogenesis process induced bybrain insult. Further studies of MI and SCI action could lead to more successful translational research and development of inositols as future anti-epileptic compounds.

Chapter 4 – Inositol (1,4,5) trisphosphate is generated at the cell surface following stimulation and via activation of its receptors on intracellular stores, is the prime regulator of cellular Ca^{2+} responses. However, Ca^{2+} regulation in the heart depends most importantly on Ca^{2+}-induced Ca^{2+} release controlled by ryanodine receptors on the sarcoplamic reticulum, raising questions about the contribution of IP_3 and its receptors (IP_3-R). Furthermore, cardiomyocytes show only limited activity in generating IP_3 and the level of expression of IP_3-R is low. However, both IP_3 generation and IP_3-R expression are heightened in many cardiac pathologies, including hypertrophy, heart failure and in atrial fibrillation. IP_3 has been suggested to contribute to disease progression by promoting arrhythmia, hypertrophy and chamber dilatation. Experimental studies have argued both for and against this proposal. Many factors are altered in diseased myocardium and the challenge is to identify those that critically influence outcomes. In the current review, the evidence that IP_3 and IP_3-R contribute to heart disease is critically re-evaluated.

Chapter 5 - Many hormone and neurotransmitter receptors utilize the phosphate-dylinositol signaling cascade. Phosphatidylinositol 4,5-bisphosphate is converted into two putative second messengers, inositol 1,4,5-trisphosphate (IP_3) and diacylglycerol. IP_3 regulates various kinds of cellular responses. It induces the release of Ca^{2+} from the intracellular stores by binding to the IP_3 receptor (IP_3R) on the endoplasmic reticulum. Many cell types, often nonexcitable, including glomerular mesangial cells (GMC) and renal vascular smooth muscle cells (VSMC), depend on this pathway to couple external signals to intracellular Ca^{2+} release. IP_3Rs not only triggers the release of calcium from intracellular stores but also opens plasma membrane calcium channels. Hepatorenal syndrome (HRS) is a common complication of advanced cirrhosis, characterized by renal failure and major disturbances in circulatory function. Renal failure is caused by intense vasoconstriction of the renal circulation.

HRS develops in approximately 55% of all patients with FHF. In order to delineate the role of inositol 1,4,5-triphosphate receptor in HRS, the authors investigated renal expression of type 1 inositol 1,4,5-trisphosphate receptors (IP_3R1) in mice with FHF. FHF model was induced by intravenous injection of lipopolysaccharide (LPS) in D-galactosamine (GalN)-sensitized mice via the caudal vein. GalN/LPS treatment led to a marked renal dysfunction and hyperkalaemia, of which kidney perfusion pressure was also significantly enhanced. However, it did not see any significant change in renal histopathology of rats in FHF. Simultaneously, the expression of protein and mRNA levels of renal IP_3R1 increased markedly, serum TNF-α and endothelial-1(ET-1) levels reached a maximum level. The authors employed an inhibitor of IP_3Rs (2-APB) in this renal failure model, and found that reduction in the glomerular filtration rate (GFR) was reversed and hyperkalaemia recovered after administration of 2-APB; whereas, 2-APB did not affect the liver injury and the expression of IP_3R1. In contrast, intravenous injection of the TNF-α monoclonal antibody into the caudal vein significantly decreased the mortality rate of GalN/LPS-treated rats from 70% to zero, and significantly improved the compromised GFR. In TNF-α monoclonal antibody treated rats, serum TNF-α, ET-1 levels and expression of IP_3R1 were decreased. The authors had demonstrated in a recent study that TNF-α could enhance IP_3R1 expression at mRNA and protein levels in human mesangial cells (HMCs), mediated through TNFR1/PC-PLC/PKCα-dependent and TNFR2-dependent IP_3R1 pathways. Isolated perfusion kidney experiments demonstrated that in normal rat kidney preperfused with TNF-α, stimulation with ET-1 could raise the perfusion pressure evidently and overexpress IP_3R1 protein and mRNA levels. Similar results were seen in GMCs and VSMCs incubated by TNF-α, the calcium release markedly increased after treated with ET-1. The findings in the present study demonstrated that TNF-α-dependent IP_3R1 up-regulation might be a specific target in future studies of renal dysfunction during the end-stage of liver disease.

Chapter 6 - Alzheimer's disease (AD) is the leading cause of adult onset dementia and the incidence is expected to rise in US, Japan and other developed countries. AD is characterized by the accumulation of amyloid–beta peptides forming plaques. Though the neuropathology of AD was discovered over a century ago, the molecular aspects of the disease development are unclear. It is speculated that amyloid-beta peptides target cholinergic neurons leading to neural degeneration. Defective calcium signaling in neurons has been shown to play a critical role in the pathogenesis of AD. There is now increased interest and research on inositol phosphates (IPs) and calcium signaling since IPs are known to regulate Ca^{2+} homeostasis in endoplasmic reticulum, cell signaling, and cell death mechanisms. Thus, dysregulation of the IP mediated calcium homeostasis and associated cell physiology seems to play a pathological role in the development of many chronic illnesses particularly neurodegenerative diseases. In addition to IPs, aluminum toxicity also has been shown to contribute significantly to the perturbance of calcium metabolism, mitochondrial dysfunction, and inflammasome activation. There appears to be a convergence of dysregulated calcium metabolism, aluminum toxicity, and inflammasome activation leading to beta amyloid deposits and associated pathology in Alzheimer's. Similar to Alzheimer's, IPs also have been shown to influence the pathogenesis of other neurological diseases including bipolar disorder. At the cellular level, there might be common shared pathways and underlying mechanisms related to IP. In this review, the pathophysiologic significance of IPs in the development of AD and bipolar disorders is discussed.

Chapter 7 - Inositol phosphates play important roles as second messengers in intracellular signal transduction. In order to study the relative affinity and specificity in binding inositol phosphates and diverse inositol phosphate-binding proteins by Surface Plasmon Resonance (SPR) analysis and pull-down analysis, biotinylated inositol phosphates, to be immobilized on avidin-based sensor chip or resin, were designed. The synthesis of the biotinylated inositol phosphates was accomplished by assembling the inositol phosphate building block, synthesized starting with optically resolved *myo*-inositol derivatives, and the biotin building block through a phosphate linkage. The inositol phosphate moiety of each biotinylated inositol phosphate showed specific binding to Pleckstrin Homology (PH) domain of PLCδ (phospholipase Cδ), PH domain of Grp1 (general receptor of phosphoinositides 1), and precursor of Gag protein ($Pr55^{Gag}$) of human immunodeficiency virus type 1 (HIV-1), as revealed by SPR analysis and pull-down analysis.

Chapter 8 - Inositol hexaphosphate (InsP6, IP6, phytic acid) is found in abundance in beans and cereals. InsP6 occurs at much lower levels in most mammalian cells, in which they regulate cellular functions. Altered inositol production has been reported from patients with diabetes mellitus, chronic renal failure, galactosemia, and multiple sclerosis.IP6 has drawn much attention due to its activity in cancer prevention and control of experimental tumor growth, progression, and metastasis. InsP6 increases differentiation of malignant cells which may then revert to the normal phenotype. IP6 treatment of all the cell lines tested so far demonstrates that it is cytostatic and not cytotoxic. InsP6 inhibits growth and induces terminal differentiation of HT-29 human colon cancer cells. Studies of the expression of tumor suppressor gene demonstrate up-regulation of wild type p53 and down-regulation of the mutant form. Its efficacy in treating central nervous system disorders including Alzheimer's disease, depression, panic disorder, and obsessive-compulsive disorder has been reported. Its usefulness as an analgesic, in pediatric respiratory depression syndrome,to mitigate adverse effects associated with lithium treatment, and in preventing neural tube defects in embryonic mice has been documented. IP(6) exhibits immunoenhancing and hypocholesterolemic activities, minimizes pathological calcification and kidney stone formation, and lowers pathological platelet activity.However,the risk of inducing uterine contractions restricts its usefulness in pregnancy.

Chapter 9 - Calcium-mobilizing receptors stimulate the hydrolysis of phosphatidylinositol 4,5-bisphosphate to 1,2-diacylglycerol and inositol 1,4,5-trisphosphate (Ins 1,4,5P3). Ins 1,4,5P3 released to the cytosol serves as a second messenger to release calcium from the endoplasmic reticulum. Growth factors such as platelet-derived growth factor, bombesin and vasopressin in fibroblasts or antigen in lymphocytes use the aforementioned transduction mechanism to produce intracellular mitogenic signals. The signaling mechanism via histamine H1-receptors is also used to mediate endothelial cell functions. Insulin regulates the phosphatidylinositol(PI) 3-kinase involving a physical association between the insulin receptor and the PI 3-kinase and tyrosine phosphorylation. Inositol is a nutrient required for keratinocyte proliferation. The Ins1,4,5P3/Ca2+ signal pathway functions to regulate such diverse processes as egg maturation and fertilization, growth, secretion, metabolism, neural activity, and excitation-contraction coupling in skeletal muscle. Myo-inositol serves as a clinically relevant osmolyte in the CNS, and its hexakisphosphate and pyrophosphorylated derivatives may play roles in such diverse cellular functions as DNA repair, nuclear RNA export and synaptic membrane trafficking. Glycerophosphoinositols induce cell proliferation in thyroid cells,modulate actin cytoskeleton organization in fibroblasts,and reduce the invasive potential of tumour cell lines. Glycerophosphoinositols promote cytokine-dependent chemotaxis in T-lymphocytes induced by SDF-1alpha-receptor activation.

In: Inositol
Editors: H. Rocha and M. Cardoso
ISBN: 978-1-62808-520-4

Chapter 1

Hypoxia- and Angiogenesis-Related *myo*-Inositol Phosphates: Synthesis and Biological Evaluation

***Alexandros E. Koumbis*[1,*] *and Konstantina C. Fylaktakidou*[2,†]**
[1]Laboratory of Organic Chemistry, Department of Chemistry, Aristotle University of Thessaloniki, Thessaloniki, Greece
[2]Department of Molecular Biology and Genetics, Democritus University of Thrace, Alexandroupolis, Greece

Abstract

One of the most fundamental physiological processes in the aerobic organisms is the delivery of oxygen to all tissues. Deprivation of oxygen leads to hypoxia which is a pathological condition playing a central role in numerous types of diseases, such as cancer (solid tumors), cardiovascular diseases and anemia. In this chapter we review the synthesis of a number of *myo*-inositol phosphates and their biological evaluation regarding hypoxia and angiogenesis in an independent or a

* E-mail: akoumbis@chem.auth.gr.
† E-mail: kfylakta@mbg.duth.gr.

combined relationship. Some of the compounds were found to be involved in the increased oxygen delivery by free hemoglobin or whole red blood cells, in the suppression of hypoxia-induced factor 1α (HIF-1α), in the inhibition of the phosphatidylinositol 3-kinase (PI3K) pathway, and in the reduction of angiogenesis. Consequently, the eradication or growth reduction of solid tumors, the prevention of sever anemia caused by sickle cell disease, and the enhanced exercise capacity of subjects with severe heart failure have been demonstrated. These results have evinced a leading role for *myo*-inositol phosphates in the treatment of diseases related to hypoxia and angiogenesis.

1. Introduction

1.1. *myo*-Inositol Phosphates

myo-Inositol phosphates is a family of mono- to hexaphosphorylated *myo*-inositols. They play a key role in biological systems as they function as secondary messengers in important cell-signaling pathways and are implicated in diverse cellular functions, such as cell growth, apoptosis, migration, and differentiation [1–3]. A number of different *myo*-inositol phosphates, which are distinguished by the number and position of the phosphate groups, were found in different organisms.

This work focuses on a variety of polyphosphorylated *myo*-inositols (with the number of phosphates ≥ 3) which have been found to be related with the phenomena of hypoxia and angiogenesis.

1.2. Angiogenesis and Hypoxia

Angiogenesis is the process of blood vessel formation. It plays a central role in a variety of physiologic events, such as organ development, differentiation during embryogenesis, growth, tissue remodeling and wound healing [4, 5]. However, angiogenesis is also involved in several pathological conditions, such as proliferative retinopathies, age-related macular degeneration, rheumatoid arthritis, cancer, and arteriosclerosis [5–8]. It consists a multi-step process, during which capillary endothelial cells (ECs) migrate and proliferate to form three-dimensional structures capable of carrying blood, i.e. the generation of a mature vasculature from a primitive vascular network occurs [4].

Among others, growth and metastasis of solid tumors depends upon the proper blood supply, which involves recruiting new blood vessels from the surrounding tissues [9]. Angiogenesis is a prerequisite for the growth of normal and drug-resistant solid tumors and is essential for their metastatic spread, as well as for the regrowth of dormant tumors [7, 10–13].

Stimulation of angiogenesis is up-regulated by tumors with the expression of a variety of angiogenic factors including fibroblast growth factors (FGFs) [14] and more importantly vascular endothelial growth factor (VEGF) [15, 16].

FGFs is a family of proteins known to possess angiogenic and lymphangiogenic abilities. They regulate a wide array of important biological phenomena such as angiogenesis, cell differentiation, tumor growth and neurogenesis [17].

VEGF is a cytokine which is considered as the most potent and specific angiogenic factor known [18]. It is secreted by tumor cells and promotes the proliferation, survival and migration of endothelial cells as well as remodeling of the neo-vessels [19]. VEGF expression is stimulated by integrin-linked kinase (ILK) via protein kinase B (Akt) and hypoxia-inducible factor 1α (HIF-1α) [20, 21] (Figure 1).

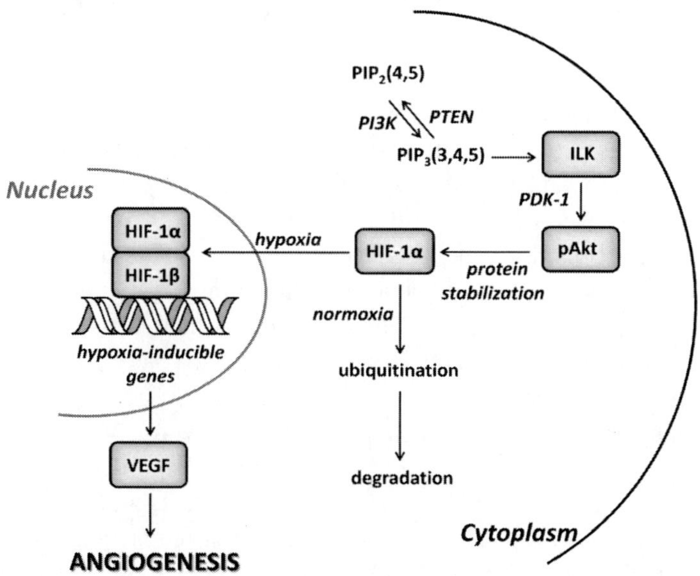

Figure 1. A simplified pathway describing how PI3K, ILK, Akt, HIFs, and VEGF are related to angiogenesis.

Thus, phosphatidylinositol 3-kinase (PI3K) catalyses the phosphorylation of inositol phospholipids PIP_2 (phosphatidylinositol -4,5-bisphosphate) to generate 3'-phosphorylated phosphoinositides PIP_3 (phosphatidylinositol-3,4,5-trisphosphate). The produced PIP_3 acts by recruiting specific signaling proteins to the plasma membrane in a mechanism mediated by their interaction with some structural protein domains. One of the best-characterized PI3K downstream targets is the serine/threonine protein kinase B (Akt) [22] and all studies concerning the role of PI3K in angiogenesis focus their attention on the PI3K/Akt pathway [23]. The crucial role of the PI3K/Akt pathway in cancer is further supported by the fact that the tumor suppressor endothelial phosphatase and tensin homolog (PTEN), whose gene is deleted or mutated in a wide variety of human cancers, possesses a 3'-phosphoinositide-phosphatase activity inactivating the PI3K/Akt pathway [24]. Inactivation of PTEN leads to permanent Akt phosphorylation that maintains tumor growth.

Regarding tumors, angiogenesis is also closely related to hypoxia. A tumor mass cannot develop above a 2 mm diameter without sufficient blood-supplied nutrients and oxygen. Therefore, hypoxia, which is a common feature of developing tumors, is one of the key conditions to turn on angiogenesis [25–27].

HIF proteins are the master transcriptional regulators of the cellular response to hypoxic tumor microenvironment [28, 29]. The increase of HIF-1α protein levels is promoted by hypoxia in a process that protects it from oxidative degradation. This event induces the transcription of VEGF and other angiogenic mediators (Figure 2) [29–31].

HIF-1α is associated with increased vessel numbers, tumor grade severity, poor prognostic, and treatment failure [32]. Hypoxia-induced tumor angiogenesis, whose purpose is to build new vessels for oxygen and nutrients supply to tumor cells, is in fact inefficient [33]. It leads to incomplete vessels that are permeable and allow metastatic spreading of tumor cells escaping through nonsealed ECs [34]. Thus, hypoxic tumor cells tent to obtain resistance to radiotherapy and chemotherapy, and are becoming highly aggressive and metastatic.

Enhancement of oxygen tension in a tumor, whose interior is hypoxic while its size increases, could therefore lead to significant inhibition of angiogenesis. Because of the oxygen-dependent activity of HIF, it represents one of the main tumor target genes in ECs for antiangiogenic strategies. According to the above, nowadays, targeting VEGF is probably one of the most attractive, promising and widely applicable antiangiogenic strategies to control and treat various malignancies.

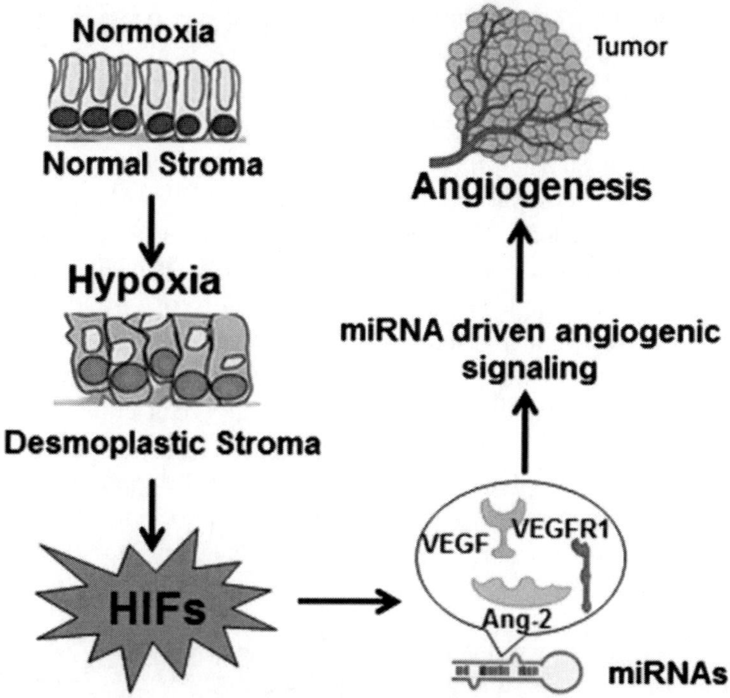

Adapted from Bao et al. *Biochimica and Biophysica Acta*, 2012 [29] (DOI: 10.1016/j.bbcan.2012.04.008) with permission from Elsevier.

Figure 2. Hypoxia driven tumor angiogenesis. Activation of HIF proteins leads to increased transcription of VEGF and this influences miRNA expression dependent angiogenic signaling.

Hypoxia is also related to other diseases. It is the most important triggering factor of painful vaso-occlusive crisis (VOC) in sickle cell disease (SCD), which is a blood disorder characterized by abnormal hemoglobin S (HbS). Under hypoxic conditions HbS polymerizes inducing distortion of red blood cells (RBCs) into rigid sickle-shaped cells [35, 36]. These form aggregates capable to interact with the endothelium, reducing, thus, blood velocity, creating local hypoxia and further enhancing sickling [37].

Finally, hypoxia is associated with heart failure, as it is known that a major determinant of maximal exercise capacity is the delivery of oxygen to exercising muscles, and to diabetes related endoneural ischemia.

2. *myo*-Inositol Hexakisphosphate

2.1. General

myo-Inositol hexakisphosphate (IHP or IP_6 or phytic acid, **1**, Figure 3) is a naturally occurring polyphosphorylated cyclitol. It is ubiquitous in plants where it usually occurs as a salt and it serves as a storage form of phosphorous. Most cereals, legumes, nuts, oil seeds, and soybean contain up to 6.4% (w/w) or even higher amounts of IHP. It is one source of dietary phosphate for both humans and animals. In mammals IHP and its lower phosphorylated analogues, produced by the hydrolytic action of several phytases, occur in almost all cell types and are important in the regulation of cellular functions and the signal transduction system [38–41]. IHP concentrations in mammalian cells could reach levels of up to 1mM [42]. The anti-nutritional aspect of IHP is well documented and occurs due to its ability to complex *in vitro* several di- and trivalent cations, and especially Ca^{2+} and Mg^{2+} *in vivo* [43].

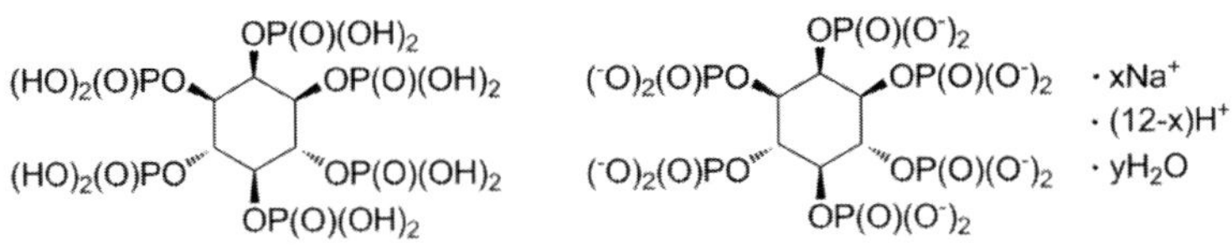

1 **2**

Figure 3. Structures of IHP (**1**) and one of its commercially available salts (**2**).

2.2. Synthesis of IHP and its Salts

IHP was first discovered by Posternak in 1903 [44]. It is a commercially available product which is usually isolated from rice and is sold as a hydrated sodium salt (**2**). The first synthesis of the perprotonated form was reported soon after its isolation [45, 46]. Thus, heating at 120-130 °C *myo*-inositol (**3**) and phosphoric acid in the presence of phosphorous pentoxide yielded IHP (Scheme 1).

Various salts of IHP have been prepared by the direct reaction of the perprotonated form with the appropriate base or from another salt with cation exchange.

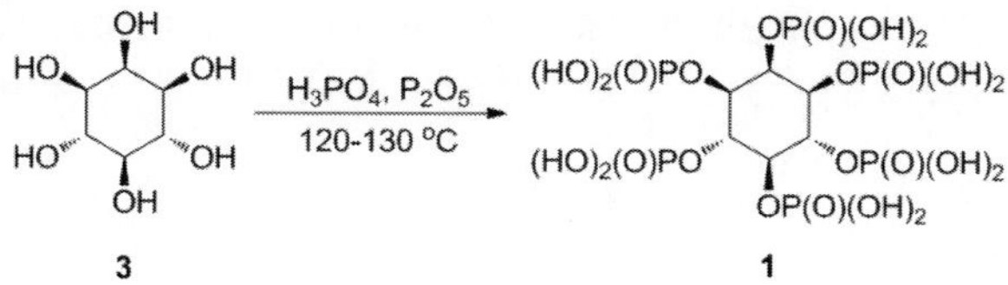

Scheme 1. Synthesis of IHP (**1**) from *myo*-inositol (**3**).

Representative examples are the hexakis Ca^{2+} **5** [47], dodekakis(triethylammonium) **6** [48], multiammonium **7** [49] salts, several mixed salts and even more complicated salts with organic counterions, such as carnitine, **8** [50] (Scheme 2).

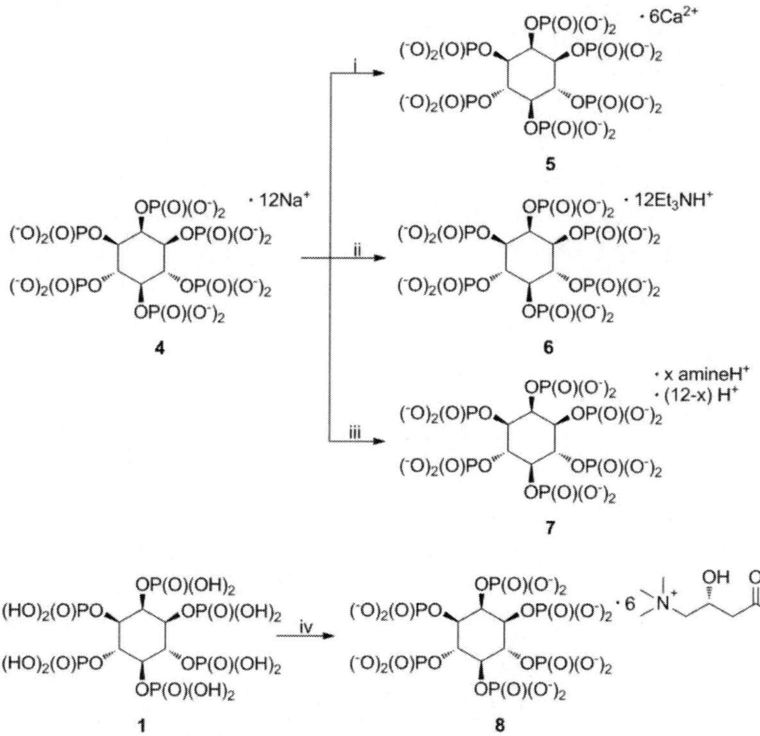

Scheme 2. Synthesis of representative IHP salts. *Reagents and conditions:* i. $CaCl_2$, H_2O, ph 8, r.t.; ii. 1) DOWEX H^+, H_2O, r.t.; 2) Et_3N, H_2O, r.t.; iii. 1) DOWEX H^+, H_2O, r.t.; 2) amine, EtOH, r.t.; iv. carnitine, H_2O, pH 4, r.t.

For simplicity reasons, in the next paragraphs, unless otherwise indicated, the abbreviation IHP represents the perprotonated form as well as its salts used in each case.

2.3. Biosynthesis of IHP

In animal cells there are two routes by which IHP (IP_6) is synthesized from I(1,4,5)P_3 (Scheme 3) [51–55]. Both of them lead to IHP via I(1,3,4,5,6)P_5. The route involving I(1,3,4)P_3 (ITPK1-dependent) was the first one discovered whereas the other more direct one (ITPK-independent) was disclosed only recently. The latter is generally believed to be the evolutionary more ancient. However, it remains unclear which of the two pathways is quantitatively more important.

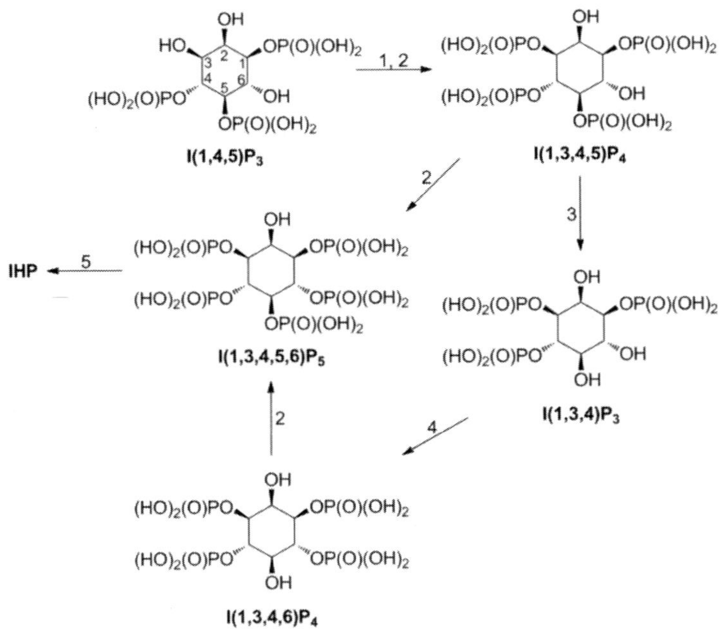

Scheme 3. The quantitatively more important routes from I(1,4,5)P_3 to IHP in animal cells *in vivo*. Numbers on arrows refer to the enzymes involved. 1: I(1,4,5)P_3 3-kinase; 2: inositol phosphate multikinase (IPMK); 3: I(1,4,5)P_3/I(1,3,4,5)P_4 5-phosphatase; 4: I(1,3,4)P_3 6-kinase/I(3,4,5,6)P_4 1-kinase (ITPK1); 5: I(1,3,4,5,6)P_5 2-kinase (IPK1).

The contribution of inositol phosphate multikinase (IPMK) to I(1,3,4,5)P_4 synthesis has been seriously questioned. Other alternative but less important routes to I(1,3,4,5,6)P_5 and consequently to IHP are also known [e.g. phosphorylation of I(3,4,5,6)P_4 by ITPK1].

For plant cells three distinct pathways for IHP biosynthesis have been proposed. Two of them are lipid-dependent since they originate from phospholipase C hydrolysis of PI(4,5)P_2 and involve the stepwise phosphorylation of I(1,4,5)P_3 and I(1,3,4)P_3 to I(1,3,4,5,6)P_5 respectively (Scheme 4) [55, 56]. The final phosphorylation to IHP is identical to the one described for animal cells. The third route (lipid-independent) is less explored. It starts probably from inositol or I(3)P and leads to I(1,3,4,5,6)P_5 through I(3,4,6)P_3 and I(3,4,5,6)P_4. Analogous biosynthetic pathways occur in yeast.

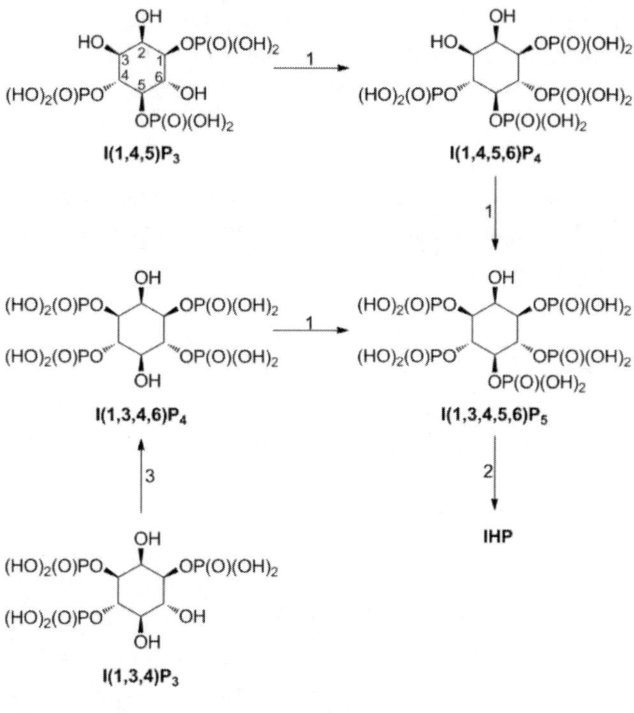

Scheme 4. The quantitatively more important routes from I(1,4,5)P_3 and I(1,3,4)P_3 to IHP in plant cells *in vivo*. Numbers on arrows refer to the enzymes involved. 1: inositol phosphate multikinase (IPMK); 2: I(1,3,4,5,6)P_5 2-kinase (IPK1); 3: I(1,3,4)P_3 6-kinase/I(3,4,5,6)P_4 1-kinase (ITPK1).

2.4. IHP and Hemoglobin

IHP is the strongest known allosteric effector of hemoglobin (Hb). It displaces hemoglobin-bound 2,3-bisphosphoglycerate (2,3-BPG), the physiological effector in mammalian RBCs since it binds to the same allosteric site 1000 times more strongly. This results in the reduction of the O_2 affinity of hemoglobin and, consequently, in the increase of the amount of O_2 released in the periphery [57, 58]. This effect is particularly important for different therapeutic applications, such as ischemic pathologies in myocardium, brain and other tissues or other hypoxia related pathological conditions such as SCD and cancer. However, IHP at physiological pH is a dissociated polyanion (it bears at least seven charges) and cannot passively permeate the erythrocyte membrane. Therefore, several research groups have tried to overcome this problem by developing different IHP encapsulating strategies. These methods involve liposomal interaction, the osmotic pulse procedure, reversed hypotonic poration, electroporation, and association with lipophilic counter cations.

Incorporation of IHP in RBCs was first achieved by Gersonde and Nicolau [59–62]. They have applied a liposomal interaction, i.e. fusion of RBCs with IHP-loaded lipid vesicles. As it was indicated by the "right-shift" of the oxygen dissociation curves (ODCs), there was a drastic, lasting and controlled decrease of the O_2 affinity of Hb and, consequently, a significant increase (280%) of O_2 release by RBCs.

An independent work by Kumpati et al. [63] using the liposomal transport system has also shown the effect of *in vitro* stimulated exchange transfusion of the IHP-loaded RBCs on sickling. These cells were found to result in higher venous pO_2 values and less sickling throughout the range of oxygen delivery. According to this concept there could be two potential mechanisms of beneficial action of IHP-loaded RBCs: a) direct oxygen transfer from IHP-loaded RBCs to sickle RBCs, and b) increased oxygen delivery to tissues lowering thus the demand for oxygen delivered by sickle RBCs.

Alternatively, the osmotic pulse method was used to obtain IHP-loaded RBCs [64]. This method subjects the cells to a short but intense period of osmotic stress promoted by DMSO. Although this procedure gave high P_{50} values (45 Torr, *ca.* 100% shift) there was a significant loss of hemoglobin (43%) from the cells decreasing the performance in terms of O_2 binding and transport. The same approach was later improved [65–67] using a continuous-flow method, in order to allow uniform treatment of large, unit-size volumes of RBCs with a relatively small quantity of reagents. Following these modified protocols the loss of hemoglobin was lowered and the post-transfusion human erythrocyte viability was increased (75% after 24 hours) whereas P_{50} increases of approximately 15 Torr were achieved.

Experiments with human and pig whole RBCs were performed by Teisseire et al. [68–70]. They have used a RBCs reversed osmotic lysis-resealing process (reversed hypotonic poration) in order to encapsulate IHP [71–73]. The main advantage of that continuous-flow procedure were large-scale transformation, control of sterility and reproducibility. Concentration-dependent P_{50}s of up to 80 Torr (160–270% shifts, Figure 4) were observed, whereas the obtained IHP-loaded pig RBCs were used *in vivo*, as well.

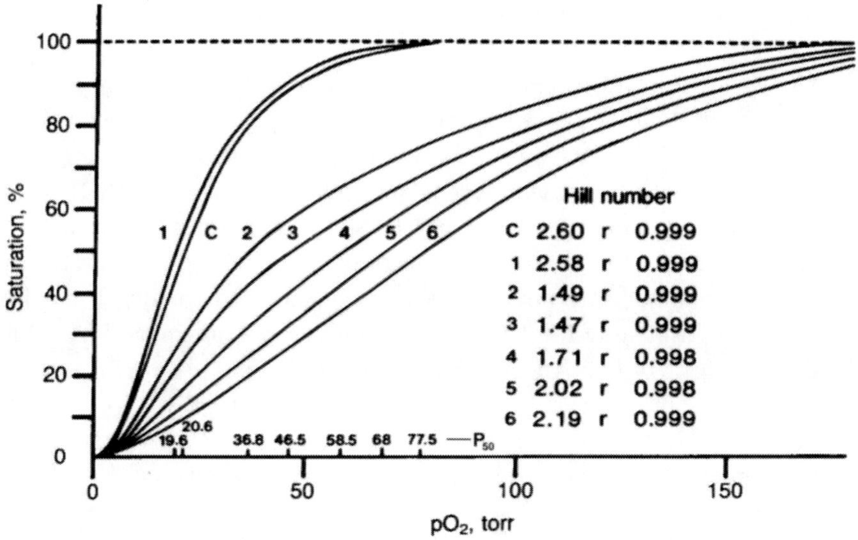

Adapted from Teisseire et al., *Proc. Natl. Acad. Sci. USA*, 1987 [74] with permission from Prof. C. Nicolau.

Figure 4. Changes in affinity obtained on same sample in pig erythrocytes (conditions: pH 7.40; PCO_2 = 40 Torr, temp. = 37 °C). Curve 1 represents erythrocytes subjected to the same procedure without IHP incorporation. Curve C is the control one. Curves 2-6 represent affinity curves drawn from erythrocytes that have been subjected to IHP incorporation procedure (various concentrations). The corresponding Hill numbers were calculated in each case from slope of a straight line between 30 and 70% saturation on log [sat/(100-sat)] vs. log pO_2 relationship. P_{50} values, calculated from each curve, are also given.

Isovolumic exchange transfusion of these high-P_{50} erythrocytes in piglets led to substantial P_{50} increases (an average of 34% shift) whereas the cardiac output was inversely decreased. This fact indicates that under physiological conditions heart reacts to the higher O_2-releasing capacity of IHP-loaded erythrocytes by reducing its output and suggests that under pathological hypoxic conditions (e.g. ischemic insult, heart stroke e.t.c.) normoxia could be restored. It is worth mentioning that life span of these IHP-loaded RBCs remained practically unchanged. Later, the authors [74] have also shown that the reduced O_2 affinity of the same IHP-loaded RBCs was still effective 20 days after transfusion in awake piglets. Furthermore, the same engineered erythrocytes were used to evaluate the influence of ODC right-ward shifts on the oxygen exchangeable fraction (OEF%). Various pO_2 to mimic normal or pathological situations were employed in order to estimate the fraction of O_2 delivered to tissues. In optimized conditions these erythrocytes were potentially able to deliver two to three times more oxygen than normal cells [75].

A similar reversed hypotonic poration method [76] was used very recently to prepare IHP-loaded mammalian RBCs. The antisickling effect of these modified erythrocytes was examined according to Kumpati's perspective [63]. Thus, the *in vitro* treatment was seven times more effective when IHP-loaded RBCs were mixed with 50% of patient RBCs [77], compared to normal stored RBCs, resulting in 35% of sickling reduction and a 94% response rate while sickling was inhibited in a dose-dependent manner. Additionally, it has been clearly shown [78] that IHP-loaded RBCs decrease RBC aggregation and blunt the adverse effects of hypoxia on RBC deformability and blood viscosity. The effectiveness of the *in vivo* treatment was assessed in two types of SCD transgenic mice [79] and resulted in an improved survival rate and brain development, prevention of severe anemia and a greatly lowered risk of VOC. This mechanism of action (see also paragraph 2.2) is depicted in Figure 5.

The potential therapeutic applications of IHP-loaded RBCs prompted Nicolau and his co-workers [80] to search for a simpler method of encapsulation of IHP in intact erythrocytes. They managed to introduce IHP in dog and mouse RBCs by static electroporation and they have observed P_{50} shifts of 100 and 66%, respectively.

An unchanged life span of these electroporated IHP erythrocytes in mice was again reported. Regarding human RBCs, an automated method of continuous-flow electroporation that is capable of yielding high volumes of low-oxygen-affinity RBCs with normal and virtually unchanged hematologic indices was reported a few years later [81]. In this *in vitro* study it was concluded that multiple pulse electroporation processes gradually increase P_{50} (100% shift after 3 pulses) without substantially affecting erythrocytes survival (more than 85%).

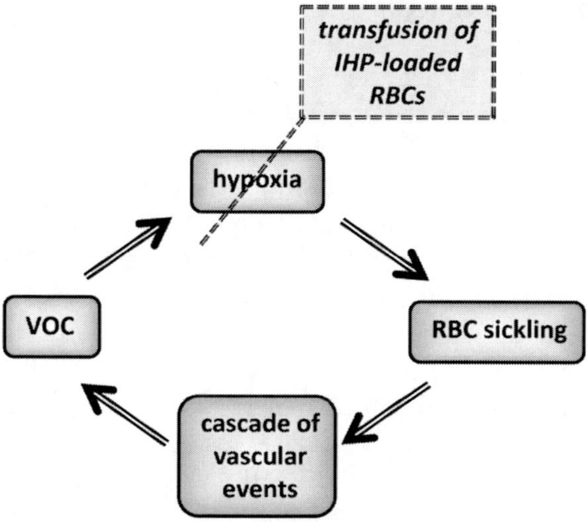

Figure 5. Using IHP-RBCs in prevention of vaso-occlusive crisis (VOC) in SCD: Transfused IHP-loaded RBCs enhance oxygenation and reduce hypoxia. This provides oxygen to tissues but also prevents sickling of abnormal RBCs and consequently VOC.

A totally different strategy based on modified IHP salts capable of passively permeating RBCs membrane was investigated by Lehn and co-workers [49]. They synthesized a library of compounds containing IHP ionically bound to lipophilic (but not lipidic) ammonium or polyammonium cations.

The ability of these water-soluble salts to act as effector of oxygen release on whole blood was determined by measuring the ODCs. In accordance to their physicochemical behavior (octanol/serum partition coefficients) cyclohexyl- and cycloheptyl-amino derivatives (**9** and **10**, respectively, Figure 6) were the best transported molecules through RBCs membrane inducing P_{50} shifts of 57% and 37%, respectively.

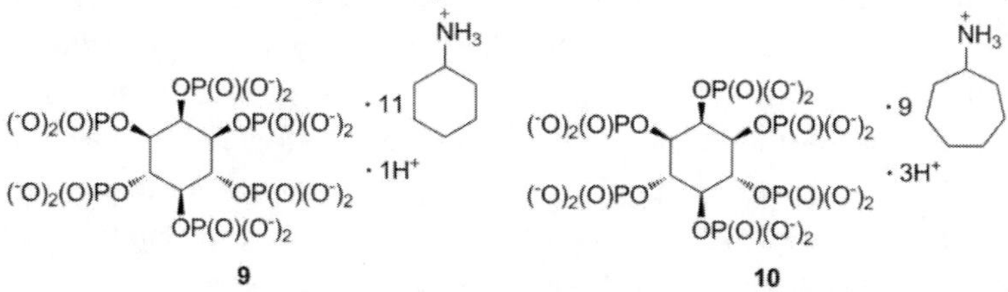

Figure 6. Structures of cyclohexyl- and cycloheptyl-ammonium IHP derivatives **9** and **10**.

2.5. IHP and Angiogenesis

Consumption of IHP-rich cereals and legumes has been shown to be associated with anticancer activity and several *in vitro* and *in vivo* studies proved its antineoplastic properties in breast, colon, liver, leukemia, sarcoma, prostate and skin cancers [38–41, 82–88]. Anticancer activity of IHP depends on its rapid intake and dephosphorylation to metabolites, mainly lower phosphorylated forms of inositol [88, 89]. Interestingly, IHP has no cytostatic or cytotoxic effect on normal cells [41].

In general, antioangiogenic activity of IHP has been associated with vascular endothelial growth factor (VEGF), fibroblast growth factors (FGFs), endothelial and inducible nitric oxide synthases (eNOS and iNOS, respectively) and matrix metalloproteinases (MMPs).

The antiangiogenic activity of IHP was initially investigated on endothelial cells [90]. Thus, it was found that proliferation of bovine aortic endothelial cells (BAECs) and differentiation of human umbilical vein endothelial cells (HUVECs) were *in vitro* inhibited in a dose-dependent manner. Reduction of basic fibroblast growth factor-induced vessel formation was shown using an *in vivo* mouse model as well. In the same study, it was found that when the human hepatoma cell line HepG2 was exposed to IHP the VEGF expression and secretion was inhibited.

Another *in vitro* study regarding the IHP influence (in combination or not with catechin) on the VEGF levels was realized using two human pancreatic cancer cells lines (PANC-1 and MIAPACA) [91]. The observed reduced VEGF levels were accompanied by increased apoptosis and reduced proliferation.

Hormone-refractory human DU145 prostate tumor xenograft growth in athymic nude mouse was used as a model to assess the *in vivo* anticancer efficacy of IHP against prostate cancer (PCA) growth [92]. It was shown that oral feeding of IHP (1 or 2% w/v in drinking water for 12 weeks) suppresses tumor growth by 40 and 66%, respectively, in terms of tumor weight/mouse. Interestingly, it was found that IHP could have pleiotropic antitumor effects targeting cell proliferation, survival, as well as angiogenesis. The latter was clearly revealed by microscopic quantification of microvessels in tumor sections using an immunohistochemical (IHC) staining of platelet endothelial cell adhesion molecule 1 (CD31). Moreover, the antiangiogenic activity was experimentally associated with an observed decrease (more than 50%) of VEGF levels.

A few years later [93] it was shown that IHP inhibits the PI3K-ILK1-Akt-mediated signaling pathway and produces strong antitumor activity in advanced and aggressive human PCA (PC-3 cell line) nude mice xenografts. This inhibition resulted to reduced proliferation and angiogenesis together with increased apoptosis. IHC staining of xenograft tissue samples of mice fed with 1 and 2% IHP doses had shown a strong decrease in ILK1 protein levels (42 and 61%, respectively) and a reduction of Akt phosphorylation (30 and 45%, respectively). A significant decrease of CD31 microvessels (35 and 57%, respectively) was also observed. The antiangiogenic response of IHP was further proved measuring the cytoplasm levels of VEGF (36 and 47% decrease, respectively) and endothelial nitric oxide synthase (eNOS) levels (49 and 72% decrease, respectively, Figure 7). Moreover, the HIF-1α DNA binding activity was also strongly reduced.

A noninvasive evaluation of IHP efficacy against prostate tumor growth and progression in transgenic adenocarcinoma of the mouse prostate (TRAMP) model was performed employing anatomical and dynamic contrast-enhanced (DCE) magnetic resonance imaging (MRI) [94]. These experiments involved male TRAMP mice which were fed with 1, 2 and 4% (w/v) IHP in drinking water and showed that suppression of tumor growth and progression occurs via altering of vascularization and without any apparent toxicity (Figure 8). IHC analysis studies were performed to relate the observed antiangiongenic activity to the inhibition of the expression of proangioangenic factors. Indeed, it was determined that VEGF, inducible nitric oxide synthase (iNOS), and phosphorylated Akt levels were all significantly reduced.

Finally, in a dose- and time-dependent *in vitro* study [95], using T24 and TCCSUP bladder cancer cell lines, VEGF levels were again significantly decreased (up to 90%) along with the reduction of tumor growth (up to 93%).

Inhibition of acidic and basic FGF binding to heparin by IHP was demonstrated by Morrison et al. [96] using spectrofluorometric methods. Later, Yu and his team [97] have shown on a molecular level the interaction of IHP with the human acidic FGF. Using various biophysical techniques (isothermal calorimetry, circular dichroism, and multidimensional NMR spectroscopy) they have elucidated the structural interactions between acidic FGF and IHP and provided the mechanistic insight as to how the latter could act as antagonist to FGF-FGF receptor signaling and possibly as an antiangiogenic agent.

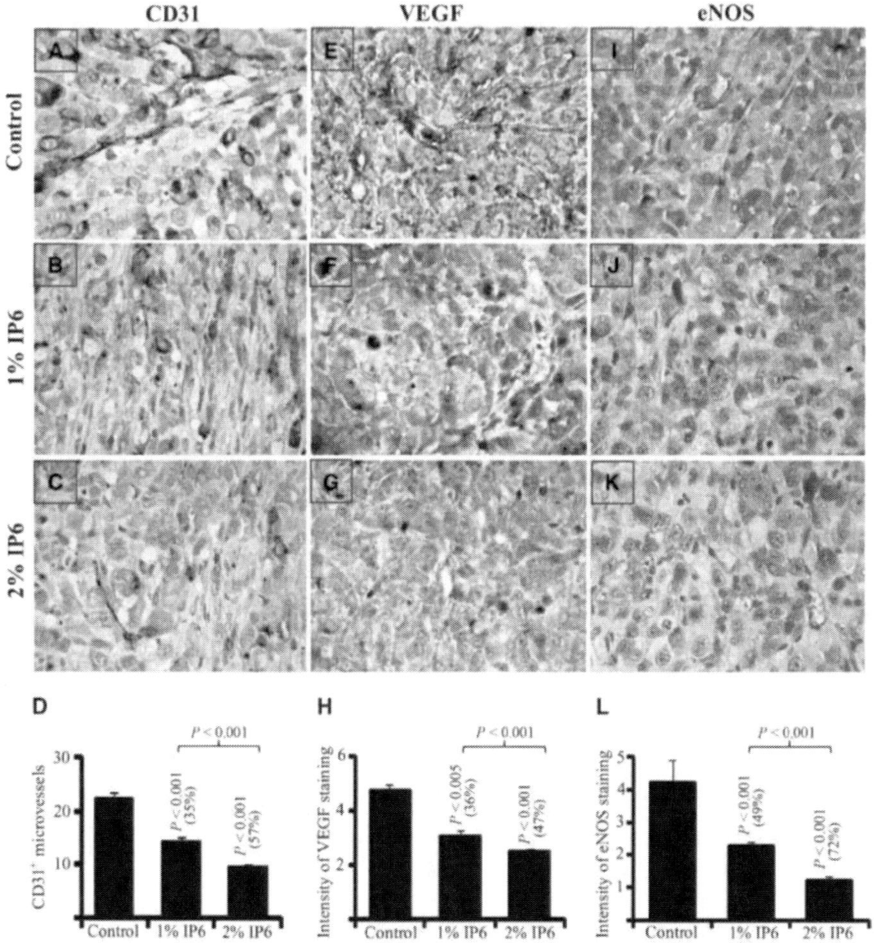

Adapted from Gu et al. *Cancer Res.*, 2009 [93] (DOI: 10.1158/0008-5472.CAN-09-2805) by permission from Prof. R. Agarwal and American Association of Cancer Research.

Figure 7. Angiogenesis inhibition and decreases of VEGF and eNOS expression in PC-3 tumor xenografts by IHP. Tissue samples from different treatment groups were subjected to IHC staining and analyzed qualitatively and quantitatively for CD31$^+$ microvessels (**A-D**) and intensity of VEGF (**E-H**) and eNOS (**I-L**) immunostaining.

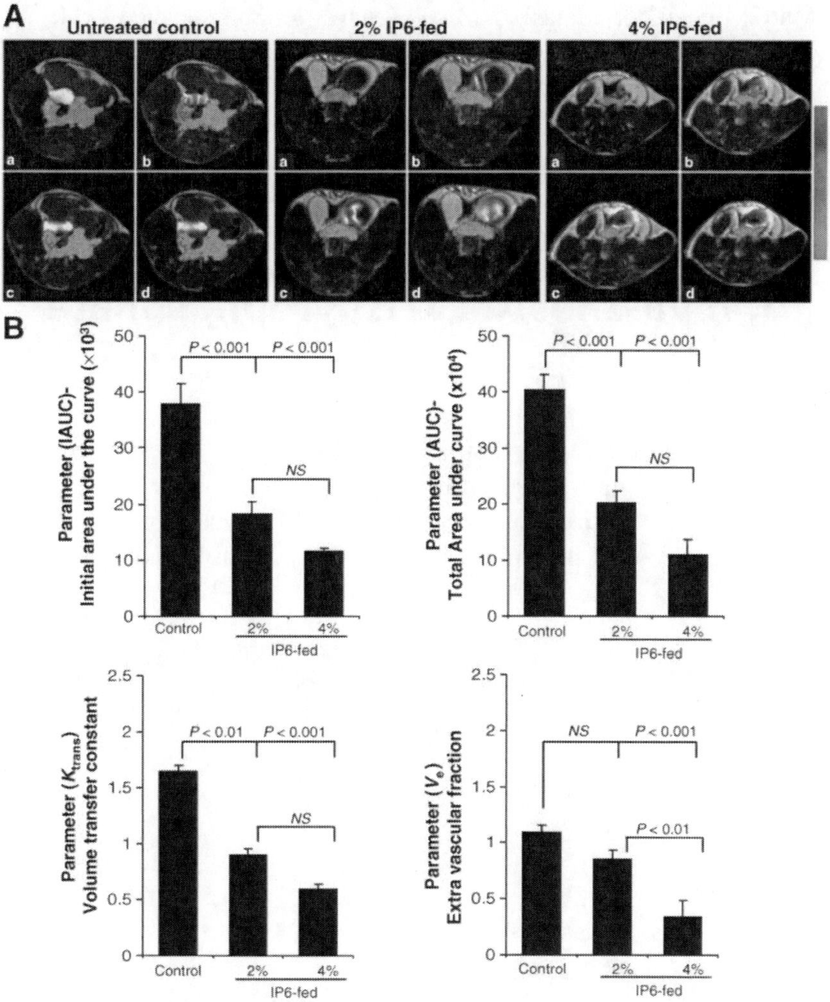

Reproduced from Raina et al. *Cancer Prev. Res.*, 2013 [94] (DOI: 10.1158/1940-6207.CAPR-12-0387) by permission from Prof. R. Agarwal and American Association of Cancer Research.

Figure 8. Dose-dependent effect of IHP feeding on angiogenesis in prostate of TRAMP mice. **A**: representative DCE-MRI images in axial plane of TRAMP prostate/ tumor. There are 4 panes per image: **a:** top left, baseline (before Gd injection); **b:** top right, 4 minutes post-gadolinium injection; **c:** bottom left, 7 minutes postgadolinium injection; and **d:** bottom right, 10 minutes post-gadolinium injection. The grey scale represents incremental increase in signal intensity in the prostate. **B:** quantitative assessment of tumor perfusion and permeability for gadolinium uptake in prostate of TRAMP mice fed with IHP.

Matrix metalloproteinases (MMPs) have long associated with tumor growth and progression, metastasis, and angiogenesis. In stimulated by phorbol-12-myristate 12 acetate (PMA) colon cancer cells (Caco-2) IHP exerts an influence of basal mRNA expression of some MMPs and their tissue inhibitors and down-regulates several MMPs [98]. These results indicated that IHP could suppress the expression of MMP genes at transcriptional level.

3. *myo*-Inositol Trispyrophosphate

3.1. General

myo-Inositol trispyrophosphate (ITPP, **11**, Figure 9) was, for several years, the only known seven-membered cyclic pyrophosphate. The identification, via spectroscopic methods, of the original structure of phytic acid was the purpose of its synthesis in 1969 [99]. It has found no further use until lately.

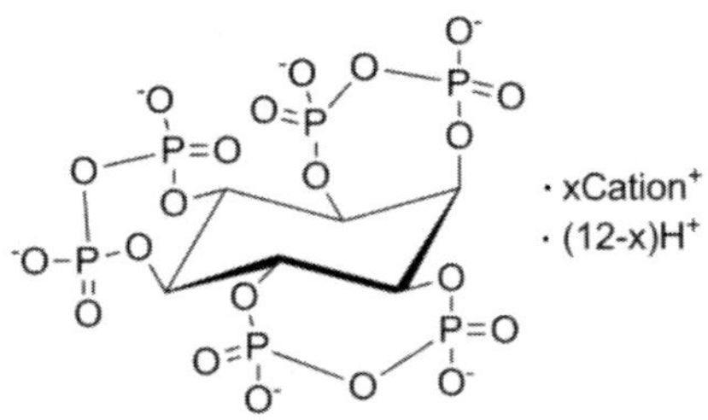

11

Figure 9. Structure of ITTP (**11**).

3.2. Synthesis of ITPP

The first reported synthesis of ITPP [99] used IHP dodecasodium salt **4**, which was transformed via the perprotonated form **1** to the pyridinium salt **12** (Scheme 5).

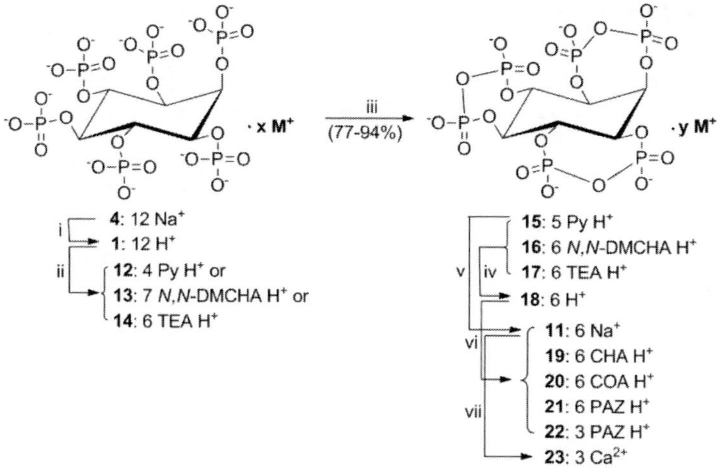

Scheme 5. Synthesis of various ITPP salts (**11**, **15-17**, **19–23**) and its perprotonated form (**18**). *Reagents and conditions:* i) DOWEX H^+, H_2O, r.t.; ii) amine, H_2O, r.t.; iii) DCC, Py/H_2O (4/1), 65 °C or DCC, MeCN/H_2O (2/1), 70 °C or DCC, EtOH/H_2O (2/1), 90 °C; iv) DOWEX H^+, H_2O, r.t.; v) DOWEX Na^+, H_2O, r.t. or NaOH, H_2O, r.t.; vi) amine, H_2O, r.t.; vii) $CaCl_2$, H_2O, r.t. [amine: Py = pyridine, *N,N*-DMCHA = *N,N*-Dimethylcyclohexylamine, TEA = triethylamine, CHA = cycloheptylamine, COA = cyclooctylamine, PAZ = piperazine].

Heating of **12** with *N,N'*-dicyclohexylcarbodiimide (DCC) in pyridine gave ITPP pyridinium salt **15**, which upon addition of NaOH and recrystallization provided the ITPP sodium salt **11**. A modification of the procedure, however, by changing the equivalents of reagents and the time of the reaction [100] led to a more efficient conversion (77% yield). Furthermore, a passage of the pyridinium salt **15** through a DOWEX Na^+ column gave directly hexasodium ITPP (**11**), in high purity and excellent yield. Accordingly, other ammonium ITPP salts (**16**, **17**, **19–22**) and the tricalcium salt **23** were synthesized via direct formation of the pyrophosphate rings or through cation exchange.

3.3. Biosynthesis of ITPP

Up to date, ITPP has not been found in nature.

3.4. ITPP and hemoglobin

Following the initial study [49] of IHP uptake by RBCs using a two phase liquid distribution system involving IHP derivatives that would be soluble in both aqueous and low polarity media, a number of novel ITPP salts [100], bearing mainly cycloalkylammonium counter cations, were investigated for their efficiency to deliver highly charged ITPP in RBCs. ITPP, being an intramolecular anhydride of IHP with considerably less negative charges, was considered to be a good choice for a prodrug approach for IHP delivery in RBCs. Other small inorganic and biologically important cations, like sodium and calcium, were also included in this study. All these lipophilic and inorganic salts of ITPP (see Scheme 5) gave P_{50} shifts of free Hb of 20–250%, whereas, remarkably, in whole blood a strong shift of *ca.* 30-40% was observed only for the inorganic ones. In the latter case the lipophilic counter cations did not seem to influence the uptake of the compound and gave a poor P_{50} shift of 7–20%. Thus, it emerged that ITPP sodium salt is a novel membrane permeant allosteric effector of hemoglobin.

It is known that the polyanionic nature of IHP and 2,3-BPG makes passive transport through the phospholipid bilayer highly unlikely without the existence of a specific transport mechanism. Remarkably, studies of ITPP sodium salt **11** with hemoglobin [101] indicated that the compound is able to enter into RBCs by a specific transport mechanism which involves the participation of the band 3 ion exchanger, a nonselective anion transporter abundant in the membranes of RBCs. Thus, when RBCs were pretreated with 4,4'-diisothiocyanatostilbene-2,2'-disulfonic acid (DIDS), an irreversible non selective inhibitor of band 3, or with NAP-taurine [*N*-(4-azido-2-nitrophenyl)-2-aminoethylsulfonate], a selective competitive inhibitor of the band 3 anion exchanger, respectively, and thereafter incubated with ITPP, no P_{50} shift has been observed in both cases. These results strongly indicated that band 3 transporter is responsible for ITPP permeation into RBCs.

3.5. ITPP and Angiogenesis

The antiangiogenic activity of ITPP was first exhibited by Kieda et al. [102]. Human microvascular endothelial cells (MECs), when subjected *in vitro* to hypoxia, align to form vessel-like structures as in the angiogenic process. This did not happen when MECs were cultured under hypoxic conditions in the presence of human ITPP-loaded RBCs, most probably because these modified RBCs are capable of releasing under hypoxia more oxygen than their "normal" counterparts (Figure 10).

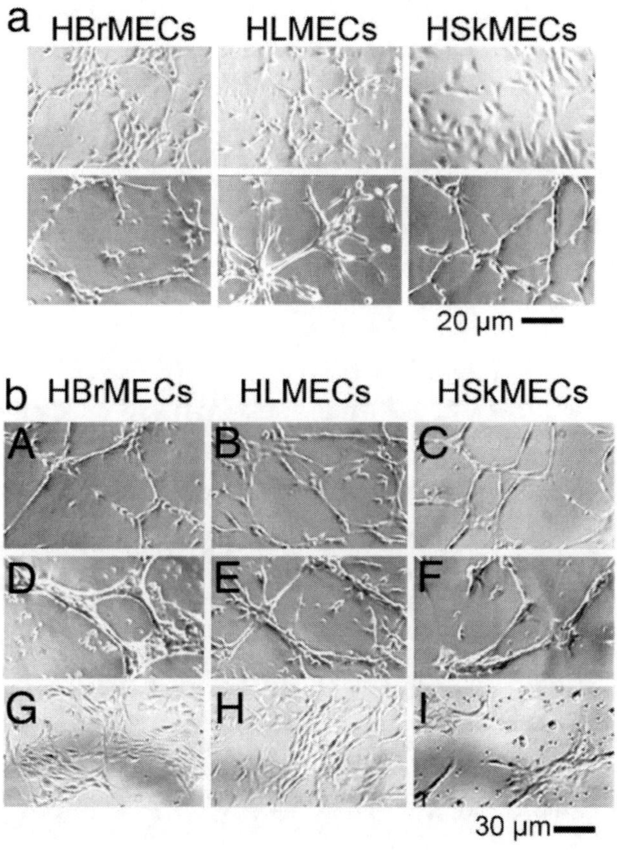

Reproduced from Kieda et al. *Proc. Natl. Acad. Sci. USA*, 2006 [102] (DOI: 10.1073/pnas.0607109103).

Figure 10. Influence of ITPP-loaded RBCs on the course of angiogenesis *in vitro* by phase contrast microscopy for human brain MECs (*a Left*, *bA*, *bD*, and *bG*), human lung MECs (*a Center*, *bB*, *bE*, and *bH*), and human skin MECs (*a Right*, *bC*, *bF*, and *bI*). (*a*) EC lines undergo faster angiogenesis under hypoxia (*Lower*) than in normoxia (*Upper*). (*b*) ECs seeded on Matrigel cultured in OptiMEM in hypoxic conditions (*bA*-*bC*), hypoxic OptiMEM with addition of normal RBCs (*bD*-*bF*), and hypoxic OptiMEM with RBCs loaded with ITPP (*bG*-*bI*).

Moreover, levels of VEGF and of HIF-1α, elevated in the human MECs under hypoxia, were dramatically reduced or even suppressed in the presence of the ITPP-loaded RBCs [102]. Treatment of these endothelial cells directly with free ITPP at different concentrations had no effect on their ability to undertake angiogenesis. This finding indicated that the enhancement of pO_2 by the increased oxygen-releasing capacity of ITPP-RBCs, which leads to higher oxygen tension in the hypoxic environment, may be linked to the observed decrease in the levels of HIF-1α and VEGF in these cells and thus to the inhibition of angiogenesis. Finally, incubation with ITPP enhanced the capacity of Hb to release bound oxygen, leading to higher oxygen tension in the hypoxic environment, thus inhibiting hypoxia-induced angiogenesis. The same effect on angiogenesis was verified with another model as well. It was found [103] that ITPP triggers defects in the vascular network of the chick chorioallantoic membrane (CAM) *in ovo*, consistent with an anti-angiogenic effect during physiological angiogenesis (Figure 11).

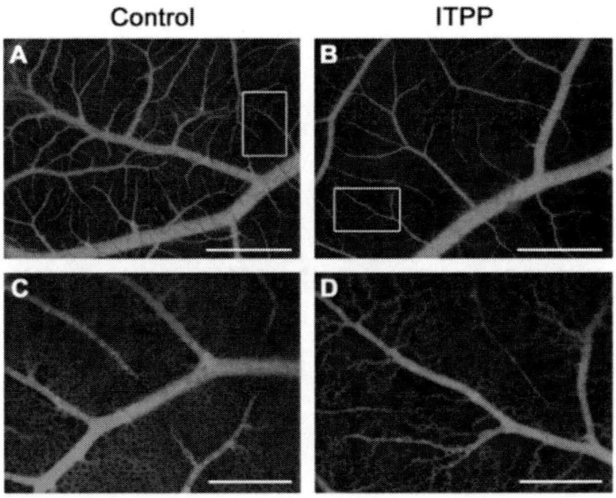

Adapted from Sihn et al. *FEBS Lett.*, 2007 [103] (DOI: 10.1016/j.febslet.2007.01.079) with permission from Elsevier.

Figure 11. ITPP defects on CAM vasculature. (*A* and *B*) Low magnification angiographic images of CAMs treated with the vehicle (A) or ITPP 0.1 M (B). (*C* and *D*) Higher magnifications of the insets of *A* and *B*, respectively. Scale bars: 2 mm (*A* and *B*) and 400 μm (*C* and *D*).

ITPP also markedly reduced tumor progression and angiogenesis in an experimental model of U87 glioma cell nodules grafted onto the CAM (Figure 12).

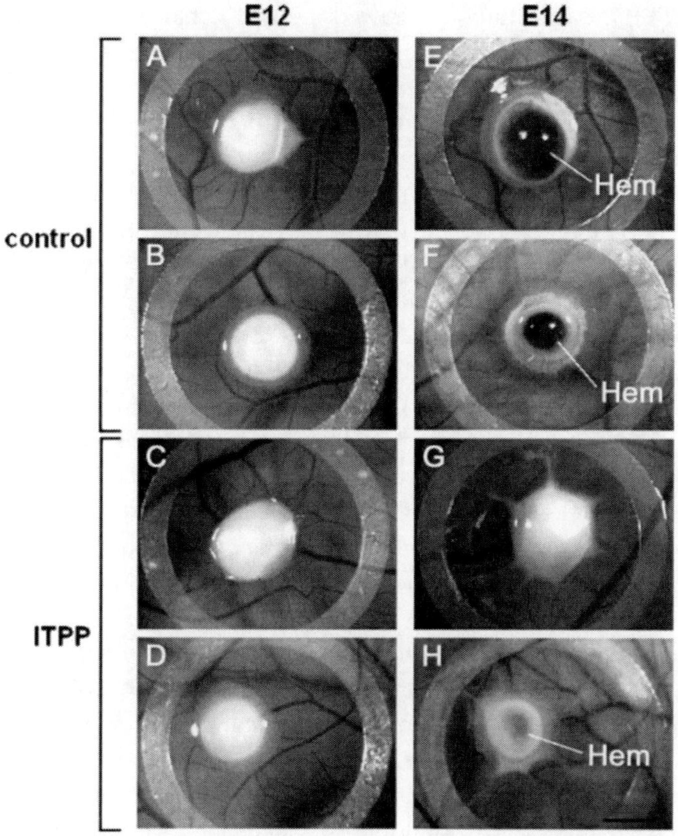

Adapted from Sihn et al. *FEBS Lett.*, 2007 [103] (DOI: 10.1016/j.febslet.2007.01.079) with permission from Elsevier.

Figure 12. Effects of ITPP on the development of experimental U87 glioma. U87 cells growing as nodules on CAM treated with vehicle (*A*, *B*, *E*, *F*) or ITPP 0.1M (*C*, *D*, *G*, *H*). (*A–D*) nodules at E12 (beginning of the treatment). (*E–F*) same nodules at E14. Hem: hemorrhages. Scale bar: 2 mm.

In a following study [104] the mechanism of action of ITPP was more deeply investigated using melanoma and breast cancer syngeneic models. ITPP treatment of those cancers stably increased oxygen tension and blood flow. It thus, produced a strong reduction of tumor growth assessed in terms of visible tumor size, its weight, and the number of tumor cells in the tumor mass. Moreover, the metastatic invasion was drastically diminished whereas the ITPP-treated animals had a survival rate of 100%.

pO_2 was measured directly inside tumors before, during, and after ITPP treatment. Non treated tumors were strongly hypoxic ($pO_2 < 2$ mmHg), whereas ITPP treated tumors showed no pO_2 change comparing to non tumoral areas. Thus, ITPP treatment prevented the formation of hypoxic areas, influencing deeply blood vessels structure, size, and density. In ITPP-treated mice, vasculature appeared less dense but organized, rather than the typical chaotic vessel architecture of a tumor.

As a consequence, ITPP suppressed hypoxia-inducible factors (HIFs) since the ITPP effect is associated with pO_2 changes in tumor. HIFs mRNA, O_2-dependent molecules like VHL, and main oxygen sensors in angiogenesis were considerably reduced or underexpressed by ITPP in treated animals. A strong enhancement of VEGF receptors 1 and 2 was observed as well.

In accordance with the finding that ITPP treatment did not influence endothelial cells growth *in vitro* [102] but controlled the angiogenic process, its *in vivo* effect was found to implicate activation of PTEN that regulates both hypoxia-dependent and hypoxia-independent mechanisms of tumor angiogenesis. As a consequence ITPP inhibited PI3K and reduced tumor Akt phosphorylation. These mechanisms normalize tumor vessels by EC reorganization, maturation, pericytes attraction, and lowering progenitor cells recruitment in the tumor.

"Normalization" accompanied a strong reduction of tumor size, and it was confirmed in terms of vessel function, first by reduction of tumor vessels permeability. ITPP-induced tumor vessel normalization was found to regulate energetic metabolism molecules linked to oxidative stress, to suppress proangiogenic/glycolysis genes and proteins cascade and to strongly reduce vascular leakage, tumor growth, drug resistance, and metastasis.

ITPP treatment avoided cancer stem-like cell selection, multidrug resistance (MDR) activation and efficiently enhanced chemotherapeutic drugs activity. As ITPP treatment improves O_2 delivery to hypoxic tissues and normalizes vessels, the authors have studied its effect on melanoma treatment by drugs such as paclitaxel and cisplatin. Combined ITPP and drug treatments acted positively and led to eradication of metastatic tumor cells from lungs. These data stress the potential of ITPP in combined therapies.

Conclusively, ITPP treatment contributed to restore vessel wall integrity and efficient blood supply by counteracting both hypoxia-dependent and hypoxia-independent HIF induction as summarized in Figure 13. ITPP action appeared to contribute in control of both mechanisms.

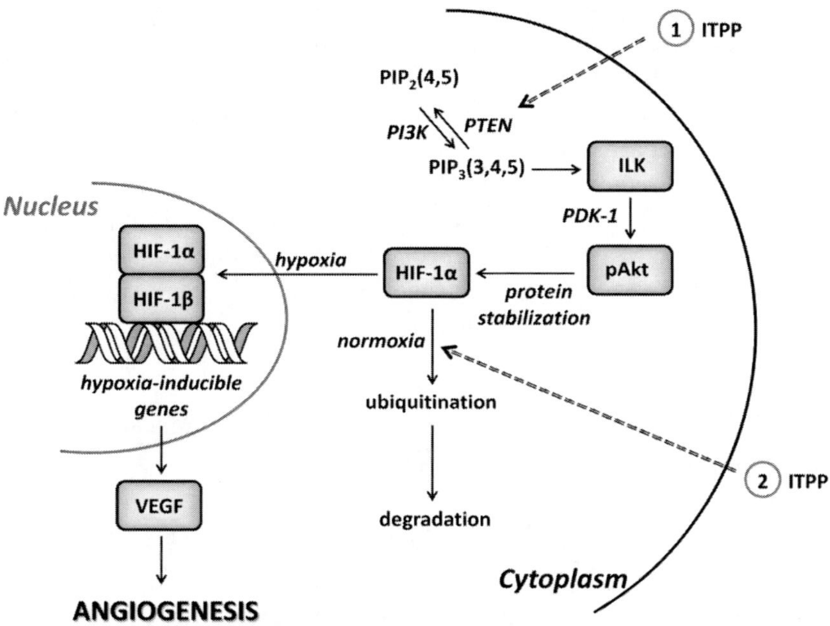

Figure 13. Proposed action of ITPP on HIFs regulation. ITPP regulates angiogenesis: **1** By activating PTEN that inhibits PI3K action and Akt phosphorylation towards HIF-1α at the endothelial cell level (hypoxia-independent mode). This regulates the vessels and increases the blood flow. **2** By acting as allosteric effector of hemoglobin. This increases the intra tumor pO_2, restoring normoxia which promotes the destabilization of HIF-1α (hypoxia-dependent mode).

The same hypothesis, that ITPP causes the suppression of HIF-1α and down-regulation of hypoxia-inducible genes and may affect tumor growth has been tested in early hepatoma tumors in rats [105]. The effect of weekly intravenous injection of ITPP on an orthotopic, syngenic rat hepatocellular carcinoma (HCC) model was compared to that for untreated animals and animals subjected to conventional Doxorubicin chemotherapy. Progression of the tumors of untreated animals resulted in full liver disease and death within seven to nine weeks. Doxorubicin delayed tumor progression, most of the animals dying between seven and twelve weeks. In contrast, within two weeks ITPP induced inhibition of tumor growth.

The treatment (terminated after four weeks) was successful, with complete tumor eradication for nine out of ten ITPP-treated animals, as demonstrated by their survival (Figure 14). The animals were monitored for a further 18 months, and no recurrence of tumors in the liver was observed. This indicated that the action of ITPP was not just growth suppression, but eradication of the tumors. In comparison, Doxorubicin was found to be dramatically less efficient: it only delayed death, with one exception. The P_{50} value of Hb in RBCs underwent a shift of 30% following ITPP injection. Under these conditions, HIF-1a activity was strongly decreased, not only in hepatic tumor but also in the surrounding liver tissue—even at some distance from the tumor, suggesting that this treatment influences the tumor's environment as well. VEGF expression was down-regulated, and apoptosis was induced in HCC and surrounding liver cells, as indicated by Caspase-3 expression. ITPP did not affect hematologic parameters during treatment.

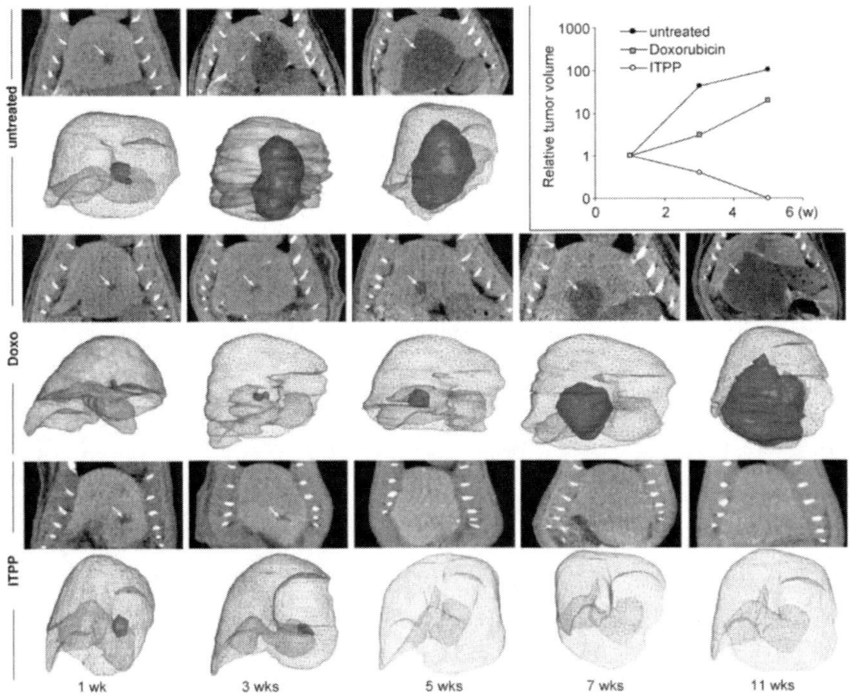

Adapted from Aprahamian et al. *ChemBioChem*, 2011 [105] (DOI: 10.1002/cbic.201000619) with permission from John Wiley and Sons.

Figure 14. Liver tumor progression as detected by mCT imaging of untreated rats (top), rats treated with Doxorubicin (center), and with ITPP (bottom). The tumor in 2D slices is indicated by an arrow (monochrome images). 3D transparent simulations created by 3D virtual-reality and rendering software (color images), from which tumor volumes were determined and plotted as a ratio to the initial volume (graph, upper right). Doxo: Doxorubicin; w: weeks.

The same has been evidenced in human colon tumors implanted in *nu/nu* mice [106]. First intraperitoneal carcinomatosis model of HT29 cells were implanted in the abdominal cavity of athymic *nu/nu* mice. Oxygen tension was monitored to assess the activity of ITPP. In untreated animals, it was 30mm Hg in muscles but only 0.2–2mm Hg in tumors, signaling their hypoxic setting; 2 h after ITPP administration, oxygen tension remained at 30–34mm Hg in muscles; however, it strongly increased up to 32mm Hg in the tumors and persisted above 25mm Hg 2 days after ITPP administration. Together, this confirmed that ITPP increased the oxygen release inside tumors. ITPP significantly reduced the growth of the subcutaneous HT29 tumors, leading to twofold reduction of the tumor volume. These results were in line with those obtained with the intraperitoneal carcinomatosis model.

To evaluate the effect of ITPP on other colon cancer cell lines than HT29, SW480 and HCT116 cells were grafted under the skin of nu/nu mice. The drug was much less efficient on HCT116 tumors. Most probably because these cells exhibit a significant constitutive expression of HIF-1α already in normoxia. Thus, ITPP treatment, which increases oxygen tension in tumors, reduces tumor growth, but not in a cell line in which HIF-1α is already expressed in normoxia.

Related to hypoxia, ITPP was found to accelerate the rise of intestine-specific homeobox gene Cdx2 expression in xenografted HT29 tumors.

The above data also stress the potential of ITPP in combined therapies. ITPP could become the adjuvant needed to chemo and radiotherapy efficacy providing enhanced O_2 supply and vessel long term normalization [107, 108] an alternative to antiangiogenic strategies [109].

The ability of ITPP to enhance the release of oxygen has prompted researchers [110] to investigate whether ITPP may have application to other diseases related to hypoxia, such as heart failure. It is known that a major determinant of maximal exercise capacity is the delivery of oxygen to exercising muscles. In normal mice, i.p. administration of ITPP (0.5–3 g/kg) caused a dose-related increase in the oxygen tension at P_{50}, with a maximal increase of 31%. After a single ITPP administration, the P_{50} increase was sustained for 48 h, had decreased by *ca.* 50% in 5 days, and was no longer present after 12 days. The prolonged duration of the effect, which is consistent with the half-life of RBCs in mice, suggests that the effect of ITTP on Hb is relatively long-lived.

In parallel experiments, ITPP caused a dose-related increase in maximal exercise capacity, with a maximal increase of 57±13%. In transgenic mice with severe heart failure caused by cardiac-specific overexpression of Gαq, i.p.

ITPP increased exercise capacity, with a maximal increase of 63±7%. To determine whether oral ingestion of ITPP causes a shift in Hb oxygen affinity, normal mice were allowed to drink *ad libitum* water in which ITPP or IHP was dissolved at a concentration of 20 mg/mL. Hb P_{50} increased by *ca.* 16% in mice that drank ITPP, but was unchanged in mice that drank water (+3%) or water with IHP (+3%), which is membrane impermeant. Thus, oral ingestion of ITPP increased maximal exercise capacity by an average of 34±10%.

Consistent with increased tissue oxygen availability, ITPP decreased HIF-1α mRNA expression in myocardium. It had no effect on myocardial contractility in isolated mouse cardiac myocytes and did not affect arterial blood pressure *in vivo* in mice. Thus, ITPP decreases the oxygen binding affinity of Hb, increases tissue oxygen delivery, and increases maximal exercise capacity in normal mice and mice with severe heart failure. These results verified that ITPP is an attractive candidate for the therapy of patients with reduced exercise capacity caused by heart failure.

4. *myo*-Inositol (1,3,4,5,6)-Pentakisphosphate

4.1. General

myo-Inositol (1,3,4,5,6)-pentakisphosphate [I(1,3,4,5,6)P_5, **24**, Figure 15], which can be isolated from avian erythrocytes, has long been known to modulate the oxygen affinity of Hb in several amphibians and the majority of birds, as does 2,3-BPG in mammals [111, 112].

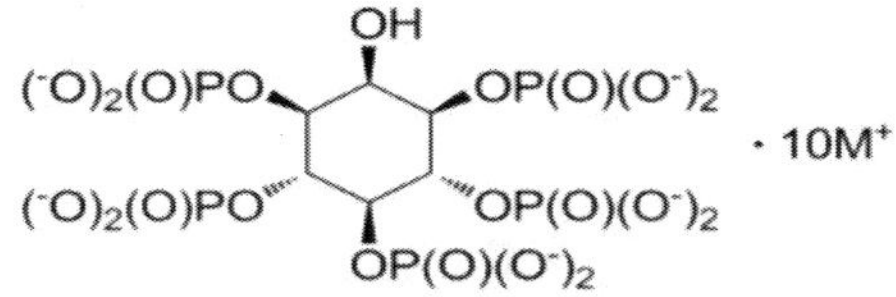

24a: M = H
24b: M = NH_3
24c: M = Na

Figure 15. Structure of I(1,3,4,5,6)P_5 (**24**).

4.2. Synthesis of I(1,3,4,5,6)P_5

A few synthetic approaches towards I(1,3,4,5,6)P_5 (**24**) have been published the last 20 years [113–116]. According to the latest method [116] synthesis began with the transesterification reaction of *myo*-inositol (**3**) with trimethyl orthobenzoate (Scheme 6). Acidic hydrolysis of the produced *myo*-inositol orthobenzoate (**25**), upon exposure to aqueous trifluoroacetic acid (TFA), provided 2-*O*-benzoylated derivative **26** in a regioselective manner. Phosphitylation of the latter using Phos (*N,N*-diethyl-1,5-dihydro-2,4,3-benzodioxaphosphepin-3-amine) followed by *m*-CPBA oxidation of the intermediate phosphite gave pentakisphosphate **27**, which upon catalytic hydrogenolysis furnished the phosphate **28**. Treatment with aqueous ammonia gave the targeted compound, **24b**, whereas cation exchange via **24a** [117] could give the corresponding sodium salt **24c**.

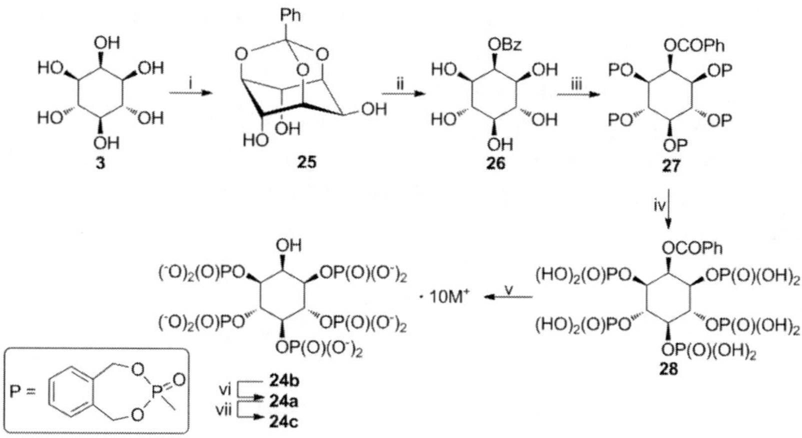

Scheme 6. Synthesis of I(1,3,4,5,6)P_5 (**24**). *Reagents and conditions*: i) $(MeO)_3CPh$, CSA, DMSO, 80 °C, 30–40 mbar, 90%; ii) TFA/H_2O (10/1), r.t., quantitative; iii) 1) Phos, DCM, 5-phenyltetrazole, r.t.; 2) *m*-CPBA, 0 °C → r.t., 96%; iv) H_2, $Pd(OH)_2/C$, MeOH/H_2O, quantitative; v) conc. aq. NH_3, 60 °C, quantitative; vi) DOWEX H^+, H_2O, r.t.; vii) NaOH, H_2O, r.t.

4.3. Biosynthesis of I(1,3,4,5,6)P_5

As it was described above (Schemes 3 and 4), *myo*-inositol (1,3,4,5,6)-pentakisphosphate represents the key phosphorylated inositol derivative for the biosynthesis of IHP in animal, plant and yeast cells.

4.4. I(1,3,4,5,6)P_5 and Angiogenesis

Maffucci et al. [117] had shown that I(1,3,4,5,6)P_5 blocks FGF-2-mediated Akt phosphorylation and inhibits both survival and migration in HUVECs. These researchers pointed out that this lack of Akt phosphorylation takes place only with I(1,3,4,5,6)P_5, whereas other polyphosphates like Ins(1,4,5,6)P_4 and IHP did not show the same effect. Moreover, I(1,3,4,5,6)P_5 inhibited the FGF-2-mediated capillary tube formation of HUVEC plated on Matrigel. Similarly the other two inositol polyphosphates used at the same concentrations had no effect. Angiogenesis *in vivo* was tested with the FGF-2-induced angiogenic reaction in BALB/c mice. Interestingly, in the presence of 50 μmol/L I(1,3,4,5,6)P_5, a clear inhibition of the FGF-2-induced angiogenic response has been observed, whereas under the same conditions, no effect was obvious in the presence of IHP, Ins(1,4,5,6)P_4, Ins(1,3,4,5)P_4, or Ins(3,4,5,6)P_4. Finally, I(1,3,4,5,6)P_5 blocked the s.c. growth of SKOV-3 xenografted in nude mice to the same extent than cisplatin and it completely inhibited Akt phosphorylation *in vivo*.

Aiming to investigate the outcome and intracellular stability of exogenously added I(1,3,4,5,6)P_5, the authors have incubated SKOV-3 with [^{32}P]I(1,3,4,5,6)P_5 for various periods. Inositol phosphates were then extracted and analyzed by high performance liquid chromatography (HPLC) to detect the different radioactive inositol compounds. They observed that I(1,3,4,5,6)P_5 was indeed able to enter SKOV-3 in a time-dependent manner and that represented a clear demonstration that, although highly negatively charged, some inositol polyphosphates can cross the plasma membrane and be internalized by cells [118, 119]. Furthermore, they have observed that the turnover of I(1,3,4,5,6)P_5 was quite slow, because only a 5.0% of the total I(1,3,4,5,6)P_5 was converted into different metabolites after 30 minutes of incubation and only a 6.2% was converted after 1 hour of incubation. The incorporated I(1,3,4,5,6)P_5 was very stable and a 84.6% of total [^{3}H]I(1,3,4,5,6)P_5 was still detectable intracellularly even after 5 hours of incubation. On the contrary, [^{3}H]Ins(1,3,4,5)P_4 was rapidly and almost completely metabolized with only a 7% of total still detectable after 5 hours of incubation [118]. These data clearly indicated that I(1,3,4,5,6)P_5 is rapidly and efficiently internalized by cells and is only minimally converted into different metabolites strongly suggesting that the observed antitumor effects were due to its activity and were not mediated by conversion to different phosphorylated compounds.

A degradation of IHP to lower phosphorylated compounds has been already observed in cancer cells although with a slow kinetic because the bulk of such compound seems to be in the hexakisphosphate form even after 6 hours of incubation [119]. Such a slow conversion of IHP to lower phosphorylated forms may easily explain the only minimal effect of IHP on Akt inhibition and, more important, the lack of effect on both the FGF-2-induced migration and tumor cell growth; that is, the amount of active compound derived from degradation of IHP is not sufficient to mediate a physiologic effect. This might also explain the requirement of very high concentrations for IHP to be active. In this respect, the authors data corroborate the reported anticancer properties of IHP that are possibly due to its conversion to I(1,3,4,5,6)P_5. Indeed, this particular IP_5 seems to be the most stable among the lower phosphorylated forms of inositol polyphosphates.

Taken together, these data indicated that specific blockade of the PI3K/Akt pathway by I(1,3,4,5,6)P_5 resulted in proapoptotic and antiangiogenic effects, and the authors showed for the first time that I(1,3,4,5,6)P_5 is able to inhibit tumor growth *in vivo*. It is noteworthy that the concentration of I(1,3,4,5,6)P_5 used in *in vivo* experiments was very low (50 mg/kg) and no toxic effects, had been detected as judged by monitoring body weight. These properties, together with the fact that I(1,3,4,5,6)P_5 is a water-soluble, natural compound, suggest that I(1,3,4,5,6)P_5 may overcome problems concerning solubility, chemical stability, and toxicity that are currently limiting the use of other potential inhibitors of the PI3K/Akt pathway.

5. *myo*-Inositol Tetrakisphosphates and Bispyrophosphates

5.1. General

The family of unsubstituted *myo*-inositol tetrakisphosphates (unsubstituted IP_4 or ITPs) comprises 15 partially phoshporylated inositol derivatives. Several of them play key roles in biological systems. They have been shown to mobilize calcium ions from internal storage [120–128], to be implicated in the regulation of calcium influx across the plasma membrane [129–137], to act as important intracellular regulators of chloride ion current in epithelial cells [138, 139], e.t.c. In contrast to ITPs, *myo*-inositol bispyrophosphates (IBPPs) have neither been found in nature nor been synthesized, with one exception [140].

5.2. Synthesis of ITPs and IBPPs

Describing the syntheses of all ITPs is out of scope of this review work. The presentation of syntheses is focused in substituted and unsubstituted 1,3,4,6-ITPs and (±)-2,3,5,6-ITPs as well as unsubstituted (±)-2,3,4,5-ITPs (general structure **29**) and the corresponding IBPPs (general structure **30**) since these derivatives were used in relevant experiments with hemoglobin (Figure 16) [140].

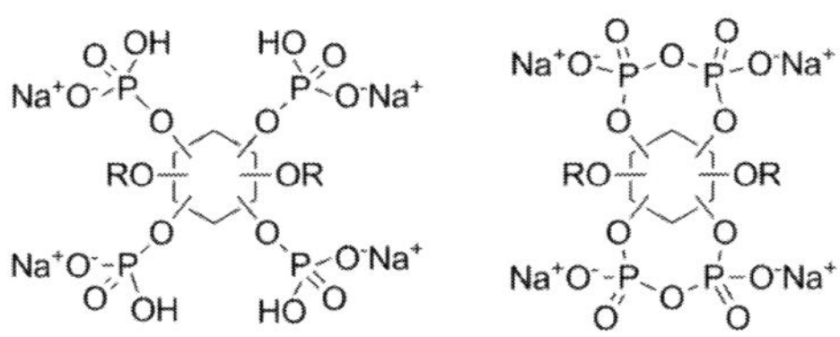

29 **30**

(R = H, Me, Et, Bu, Ac, Bt)

Figure 16. General structures of targeted ITPs (**29**) and IBPPs (**30**) in reference [140].

Various salts (or the perprotonated forms) of unsubstituted 1,3,4,6-ITP [I(1,3,4,6)P_4], 2,3,5,6-ITP [I(2,3,5,6)P_4] and 2,3,4,5-ITP [I(2,3,4,5)P_4] have been reported previously [120, 121, 141–151], whereas the preparation of their unsubstituted as well as substituted tetrasodium salts was reported recently [140]. For the symmetrical 1,3,4,6-ITPs the *myo*-inositol protected derivative **31** (Scheme 7) was used as starting material. It gave after alkylation or acylation and subsequently deprotection and perphosphorylation the key intermediates **34**. These were globally debenzylated upon catalytic hydrogenolysis to give in one step the targeted ITPs **35**. Phosphates **34** were also used, following a direct protocol or an indirect approach (via **35**), to obtain tetrakis(triethylammonium) salts **36** which served as the precursors of 1,6:3,4-IBPPs **37**. Formation of the pyrophosphate rings was achieved upon heating in the presence of DCC of **36**. The tetrasodium salts **37** were obtained after cation exchange.

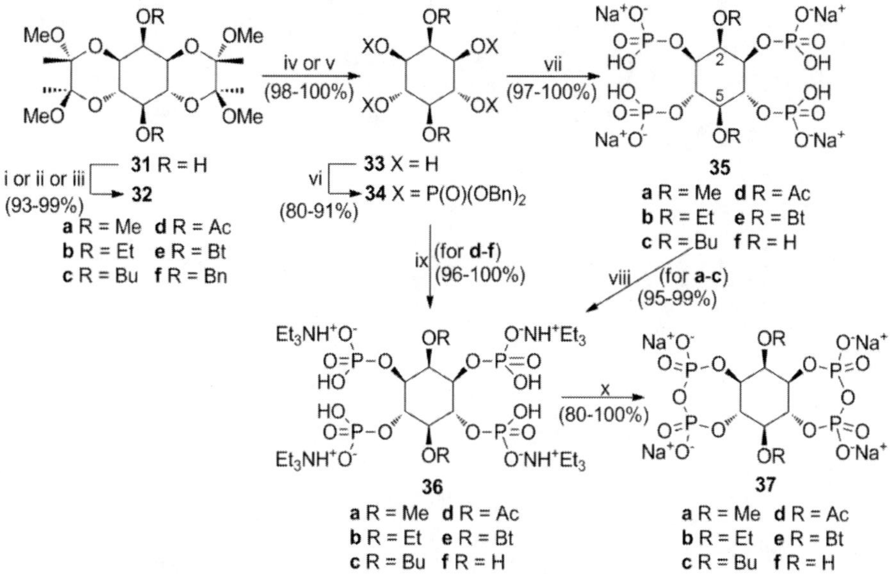

Scheme 7. Synthesis of 1,3,4,6-ITPs **35** and 1,6:3,4-IBPPs **37**. *Reagents and conditions*: i) NaH, RI, DMF, 0 °C→r.t. (for **a**, **b**, **c**); ii) R_2O, Et_3N, DMAP, DMF, DCM, 0 °C → r.t. (for **d**, **e**); iii) NaH, BnBr, DMF, 0 °C→40 °C (for **f**); iv) TFA, H_2O, r.t. (for **a**, **b**, **c**); v) TFA, DCM, H_2O, r.t. (for **d**, **e**, **f**); vi) 1) $(BnO)_2PN(iPr)_2$, 1*H*-tetrazole, MeCN, r.t.; 2) *m*-CPBA, DCM, −40 → 0 °C; vii) Pd/C, H_2 (1 or 3 Atm), $NaHCO_3$, EtOH/H_2O (1/1), r.t.; viii) 1) DOWEX H^+, H_2O, r.t.; 2) Et_3N, H_2O, r.t.; ix) Pd/C, H_2 (1 or 3 Atm), Et_3N, EtOH/H_2O (1/1), r.t.; x) 1) DCC, MeCN/H_2O (2/1), 70 °C; 2) DOWEX H^+, H_2O, r.t.; 3) DOWEX Na^+, H_2O, r.t.

For the racemic 2,3,5,6-ITP series an analogous synthetic scheme was realized (Scheme 8). Thus, the *myo*-inositol protected derivative **38** gave after alkylation or acylation and subsequently deprotection and perphosphorylation the key intermediates **41**. Catalytic hydrogenolysis promoted debenzylation and delivered the targeted ITPs **42**. Phosphates **41** were again used, following a direct protocol or an indirect approach (via **42**), to reach tetrakis(triethylammonium) salts **43**. 2,3:5,6-IBPPs **44** were obtained after DCC-promoted pyrophosphate ring formation and cation exchange. The racemic 2,3,4,5-ITP **49** was prepared following an identical synthetic scheme (Scheme 9), using this time as starting material the biscyclohexylidene *myo*-inositol derivative **45** via phosphate **48.**

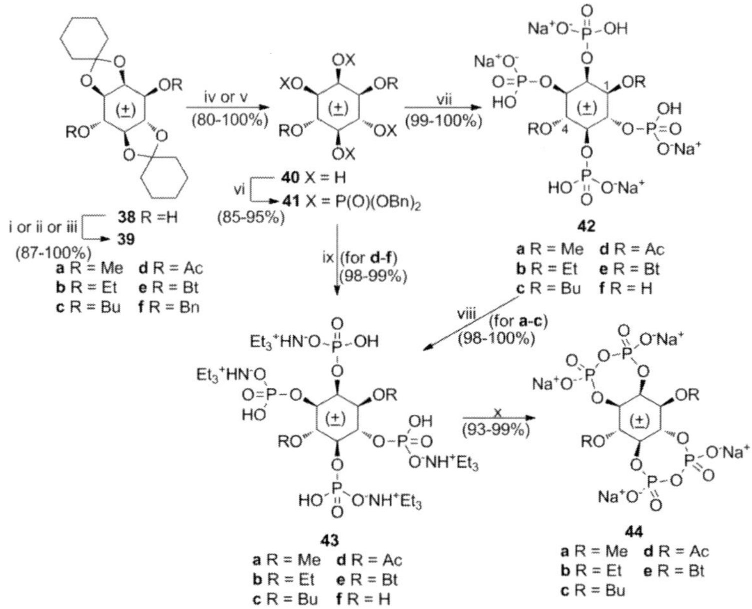

Scheme 8. Synthesis of 2,3,5,6-ITPs **42** and 2,3:5,6-IBPPs **44**. *Reagents and conditions*: i) NaH, RI, DMF, 0 °C→r.t. (for **a**, **b**, **c**); ii) R_2O, Et_3N, DMAP, DMF, DCM, 0 °C → r.t. (for **d**, **e**); iii) NaH, BnBr, DMF, 0 °C→40 °C (for **f**); iv) CSA, DCM/MeOH (2/1), r.t. (for **a**, **b**, **c**, **f**); v) TFA, DCM, H_2O, r.t. (for **d**, **e**); vi) 1) $(BnO)_2PN(iPr)_2$, 1*H*-tetrazole, MeCN, r.t.; 2) *m*-CPBA, DCM, −40 → 0 °C; vii) Pd/C, H_2 (1 or 3 Atm), $NaHCO_3$, EtOH/H_2O (1/1), r.t.; viii) 1) DOWEX H^+, H_2O, r.t.; 2) Et_3N, H_2O, r.t.; ix) Pd/C, H_2 (1 or 3 Atm), Et_3N, EtOH/H_2O 1/1, r.t.; x) 1) DCC, MeCN/H_2O (2/1), 70 °C; 2) DOWEX H^+, H_2O, r.t.; 3) DOWEX Na^+, H_2O, r.t.

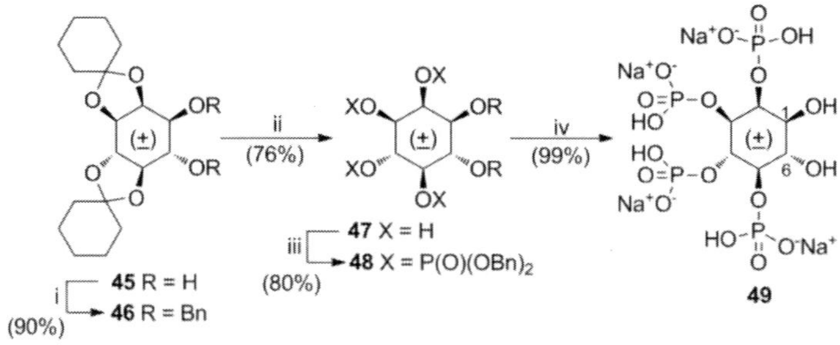

Scheme 9. Synthesis of 2,3,4,5-ITP **49**. *Reagents and conditions*: i) NaH, BnBr, DMF, 0 → 40 °C; ii) AcCl, DCM/MeOH (1/1), r.t.; iii) 1) $(BnO)_2PN(iPr)_2$, 1*H*-tetrazole, MeCN, r.t.; 2) *m*-CPBA, DCM, −40 °C→0 °C; iv) Pd/C, H_2 (1 Atm), $NaHCO_3$, EtOH/H_2O (1/1), r.t.

5.3. Biosynthesis of ITPs

Whereas I(1,3,4,5)P_4, I(1,3,4,6)P_4 and I(1,4,5,6)P_4 are known key intermediates in the biosynthetic routes of IHP (see Schemes 3 and 4), it is relatively less studied whether other unsubstituted ITPs, such as I(2,3,5,6)P_4 / I(1,2,4,5)P_4 and I(2,3,4,5)P_4 / I(1,2,5,6)P_4 exist in certain types of cells and how they originate. It has been shown that certain phytases are capable to dephosphorylate IHP and produce ITPs. The main phytate degradation route (elucidated in legumes, cereals, *Saccharomyces cerevisiae* and *Pseudomonas*) starts with the production of two IP_5: I(1,2,4,5,6)P_5 and I(1,2,3,5,6)P_5 upon the action of 4-phytases (*myo*-inositol hexakisphosphate 6-phosphorylase) on IHP. Then, further dephosphorylation of either of these pentakisphosphates leads to I(1,2,5,6)P_4 [152–154]. In other cases, as it was found for *Klebsiella terrigena* phytase [155], I(1,2,3,5,6)P_5 gives I(2,3,5,6)P_4 via an alternative less preferable route. In contrast, up to date, IBPPs have not been found in nature.

5.4. ITPs, IBPPs and Hemoglobin

The search for novel, less polar than IHP phosphates, allosteric effectors of hemoglobin had prompted Lehn and his co-workers [140] to synthesize and evaluate ITPs **35**, **42** and **49** as well as IBPPs **37** and **44** (Schemes 7, 8, and 9) for their capability to induce oxygen release from stripped human hemoglobin (Hb). Most of these phosphates and pyrophosphates proved to be efficient allosteric effectors, with similar affinities to that of IHP. The efficacy was found to be higher for free phosphates (P_{50} shifts: 144-382%) than pyrophosphates (P_{50} shifts: 11-89%). As allosteric Hb effectors these compounds enable enhanced oxygen release. The binding of these molecules to human Hb and their effects on oxygen release were investigated in order to gain insight into molecular recognition features and structure-activity relationships (SAR) in allosteric Hb regulation. It was found that these effects increase with the strength of Hb binding and correspond primarily to electrostatic interactions, i.e. the number of negative charges present, as it was also shown for other polyphophosphates [156, 157]. This trend was observed for ITPs and IBPPs and is in accordance with the comparison between ITPs and IHP as well as IBPPs and ITPP (Figure 17). Stereochemical and steric factors also play a significant but secondary role in molecular recognition. According to these results a molecular recognition pattern of the tested compounds in the allosteric pocket of Hb, indicating the crucial pharmacophoric moieties within the *myo*-inositol tetrakisphosphate scaffold, was proposed (Figure 18).

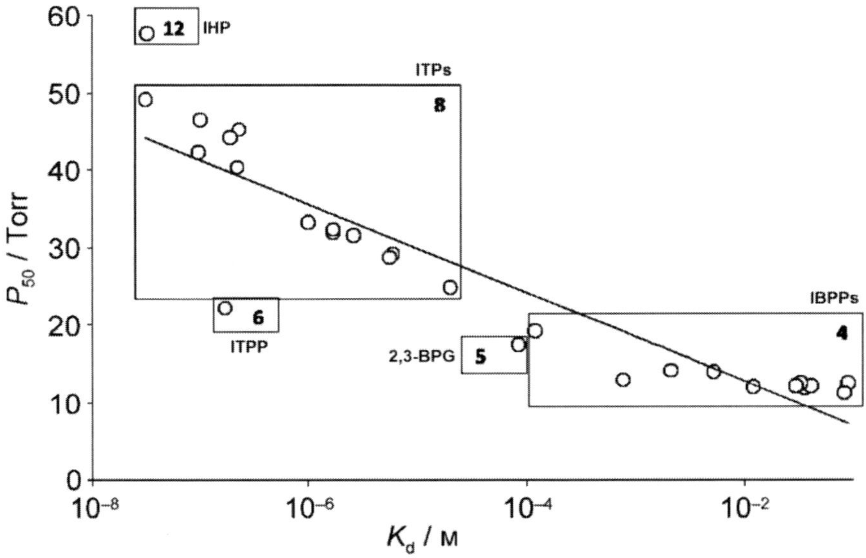

Adapted from Koumbis et al., *ChemMedChem*, 2011 [140] (DOI: 10.1002/cmdc.201000421) with permission from John Wiley and Sons.

Figure 17. Relationship between Hb-oxygen binding (P_{50}) and Hb dissociation constants (K_d) for compounds 2,3-BPG, IHP (**4**), ITPP (**11**), ITPs (**35**, **42** and **49**), and IBPPs (**37** and **44**). The line corresponds to the linear regression function ($R^2 = 0.841$) for all compounds studied. The rectangles collect groups of compounds with the same number of negative charges (in their fully ionized state), shown within each rectangle in bold.

6. *myo*-Inositol (1,2,6)-Trisphosphate

6.1. General

myo-Inositol (1,2,6)-trisphosphate [I(1,2,6)P_3 or α-trinositol, **50**, Figure 19] is a regioisomer of the well known second messenger I(1,4,5)P_3 (**51**).

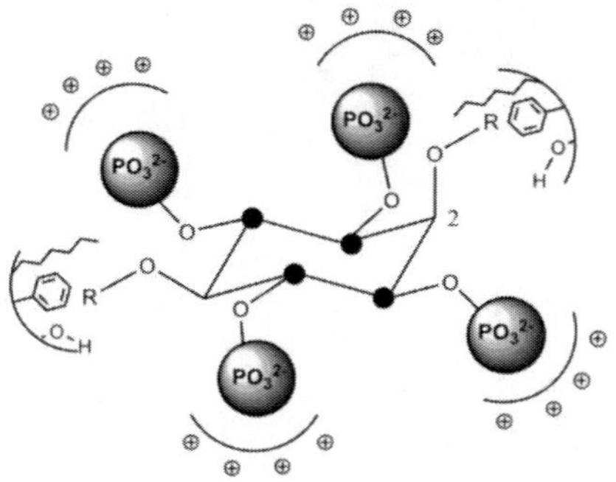

35

Adapted from Koumbis et al., *ChemMedChem*, 2011 [140] (DOI: 10.1002/cmdc.201000421) with permission from John Wiley and Sons.

Figure 18. Proposed molecular recognition of most active ITPs (**35**) in the allosteric pocket of Hb, showing essential pharmacophoric moieties within the *myo*-inositol tetrakisphosphate scaffold. Black dots indicate the positions of phosphate groups that interact with a positively charged domain of the allosteric pocket.

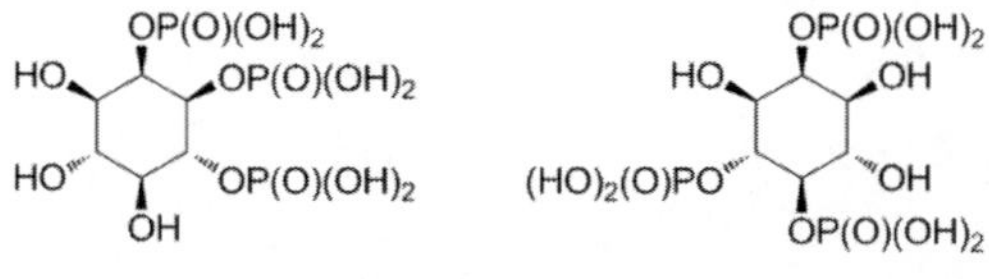

Figure 19. Structures of $I(1,2,6)P_3$ (**50**), and $I(1,4,5)P_3$ (**51**).

6.2. Synthesis of $I(1,2,6)P_3$

Baker's yeast contains phytases which selectively hydrolyze IHP phosphate groups. Fermentation of IHP (**1**) using baker's yeast (Scheme 10) is used to prepare α-trinositol [158, 159]. This enzymatic procedure can be run on a kilogram scale [160] and after ion exchange chromatography produces highly optically pure α-trinositol (*ee* > 99.7%). Alternative stereo- and enantioselective routes to α-trinositol from *myo*-inositol involve a combination of enzymatic and chemical reaction steps [161] or just a multi-step reaction sequence [162].

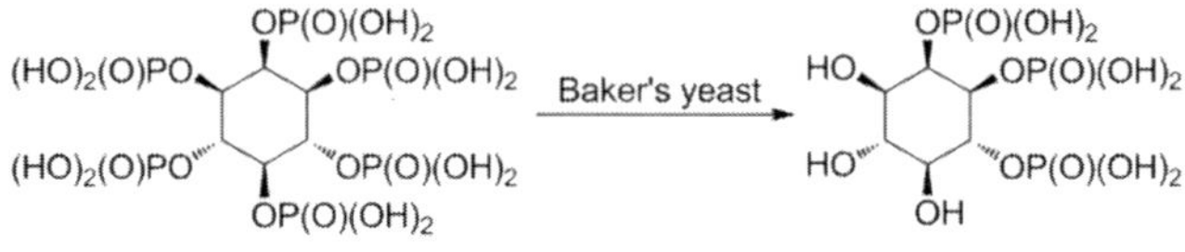

Scheme 10. Enzymatic synthesis of α-trinositol (**50**) from IHP (**1**).

6.3. Biosynthesis of I(1,2,6)P_3

Phytases (IHP phosphohydrolases) catalyze the hydrolysis of IHP into inositol pentakis-, tetrakis-, tris-, bis-, and mono-phosphates. This activity has been investigated in various plants and microorganisms [154, 163–169]. Different phytases may be involved in different IHP degradation pathways and lead, thus, to the generation and accumulation of different *myo*-inositol phosphate intermediates. In several cases, the initially formed pentakisphosphate [I(1,2,3,5,6)P_5 or I(1,2,4,5,6)P_5] is further degraded to I(1,2,6)P_3 and this represents the main dephosphorylation pathway.

6.4. I(1,2,6)P_3 and Hypoxia

Much evidence favors the development of endoneurial ischaemia in diabetes as a major factor promoting increased resistance to acute ischaemic/hypoxic conduction block. Non-diabetic rats, kept in an atmosphere containing 10% oxygen and consequently sustaining persistent central hypoxaemia, also develop conduction defects similar to those seen in normoxic diabetic rats [170]. Thus, drugs which might antagonize endoneurial vasoconstriction or increase nutritive blood flow in nerve by any other mechanism, are of interest for their insight into pathogenesis as well as their possible value in management.

I(1,2,6)P_3 has been studied for its ameliorating effect in the etiology of diabetic nerve dysfunction [171]. Those experiments were designed to determine the effects of treatment with I(1,2,6)P_3 (**50**) or parent *myo*-inositol (**3**) to diabetic rats, on nerve dysfunction.

Three weeks of experimental diabetes in the rat caused a significant decrease in motor nerve conduction velocity measured *in vivo* (45.3±3.6 m/s; mean±S.D.) compared to controls (57.7±4.5 m/s). *myo*-Inositol administration to diabetic rats (500 mg/rat per day) for the duration of the study, partially prevented this decrease (50±4.4 m/s). I(1,2,6)P_3 at a dose of 24 mg/rat per day completely prevented this reduction in diabetic rats (53.4±5.8 m/s). Resistance to hypoxic conduction block was determined *in vitro* in endoneurial preparations and was assessed as the decline in compound action potential amplitude over a 40 min period of hypoxia. Compound action potential amplitude (as % of initial value ± S.D.) was significantly greater in diabetic preparations compared with controls at 40 min of hypoxia (76.1±9.1 vs. 54.8±14.7 respectively). Treatment to diabetic rats with *myo*-inositol did not significantly affect this value (79.9±16.6) but $I(1,2,6)P_3$ treatment partially prevented the increased resistance to hypoxic conduction block (69.4±16.0).

Therefore this study demonstrated that acute abnormalities of nerve function in early experimental diabetes may be attenuated by the administration of $I(1,2,6)P_3$, possibly acting via a vascular mechanism.

Conclusion

IHP is a natural product and the best allosteric effector of Hb. However, its polyanionic nature makes difficult its transportation through the RBCs membrane. Therefore, a number of encapsulating procedures (e.g. liposomal interaction, osmotic pulse method, reversed hypotonic poration, electroporation, and association with lipophilic counter cations) have been applied in order to obtain IHP-loaded RBCs with increased ability to release O_2 and normal viability. Some of these procedures have been further evolved to continuous-flow processes. The *in vitro* and, in some cases, *in vivo* ability of these modified RBCs to counteract hypoxia has been well documented whereas the antisickling effect was successfully examined *in vitro* as well as *in vivo* (transfusion) experiments. Additionally, the anti-angiongenic activity of IHP has been associated with VEGF, FGFs, eNOS, iNOS, and MMPs using cell lines or feeding protocols. These experiments (most of them monitoring the levels of VEGF) involved cancer cells (prostate, pancreatic, bladder and colon) or normal ECs and has shown that, among others, IHP is capable of reducing angiogenesis both *in vitro* and *in vivo*. The potential therapeutic applications of IHP are obvious, considering, however, its well known anti-nutritional properties.

ITPP is a compound produced from IHP in a straightforward manner. As an overview of the studies up to date concerning ITPP we briefly note that ITPP inhibits angiogenesis in endothelial cells and in a chorioallantoid membrane model. It leads to oxygenation of the tumor interior and its surrounding areas and to normalization and maturation of tumor vasculature. It suppresses HIF-1α and expression of hypoxia genes in *in vitro* and *in vivo* models. In cancer models it reduces the growth of xenografted glioma and it leads to the eradication of early liver tumors. It also causes reduction of the tumor growth and lung metastasis in melanoma, as well as colon cancer. ITPP enhances chemotherapeutic drug's activity and avoids cancer stem-like cell selection. Finally, it ameliorates the exercise capacity of transgenic mice with severe heart failure. The above seem to be the result of two distinct mechanisms which have both been carefully investigated; a) ITPP, as an allosteric effector of Hb, enhances the capacity of Hb to release bound oxygen, leading to higher oxygen tension in the hypoxic environment and b) ITPP selectively activates the tumor suppressor phosphatase and tensin homolog (PTEN) *in vitro* and *in vivo* at the EC level thus inhibiting PI3K, reducing tumor Akt phosphorylation. Collectively, these findings highlight the multifactor and potent therapeutic use of ITPP and demonstrate its fundamental interest for advancing therapy of hypoxia and angiogenesis-dependent pathologies and suggest a significant clinical potential for ITPP in cancer therapy.

Regarding the synthesis of the less investigated lower phosphates, both I(1,3,4,5,6)P_5 and the ITPs/IBPPs libraries were prepared from parent *myo*-inositol, whereas I(1,2,6)P_3 was prepared by the phytase-promoted degradation of IHP. The *in vitro* and *in vivo* experimental data collected for I(1,3,4,5,6)P_5 strongly suggests that this pentakisphosphate blocks the PI3K/Akt pathway resulting in proapoptotic and anti-angiogenic effects and inhibition of tumor growth. On the other hand, the hypoxia-promoted acute abnormalities of nerve function in experimentally diabetic rats have been attenuated by the administration of I(1,2,6)P_3. The *in vitro* ability of ITPs and IBPPs to enhance the O_2 release from RBCs was also demonstrated with some of the ITPs showing similar affinities to Hb to that of IHP. Interestingly, the number of compounds used made easier to establish a relation between the number of charges and the strength of Hb binding (analogous increase). ITPs may represent a new generation of lower phosphates, IHP analogues, capable of permeating RBCs membrane.

In general, this review work, presents the key-syntheses, biosyntheses, and biological evaluation, regarding hypoxia and angiongenesis, of a number of *myo*-inositol phosphates. Most of these compounds exhibit significant activity, either counteracting hypoxia or/and inhibiting angiogenesis. We believe that this is an indeed fertile research field both synthetically and biologically and expect that future works will further clarify the activity of the presented phosphates as well as discover new ones. It is worth mentioning that ITPP has demonstrated safety in Phase 1 human clinical trials and is ready to begin the next phase of clinical trials on cancer patients (http://normoxys.com/about-us).

Acknowledgment

The authors would like to thank Professors J.-M. Lehn and C. Nicolau for their inspirational initiation on this research field and NormOxys Inc. (Wellesley, MA, USA) for funding the related research projects in which the authors have been involved.

Abbreviations

Ac: acetyl;
Akt: protein kinase B;
BAEC: bovine aortic endothelial cell;
BPG: bisphosphoglycerate;
Bt: butyryl;
Bu: butyl;
CAM: chorioallantoic membrane;
CD31: platelet endothelial cell adhesion molecule 1;
CHA: cyclohexylamine;
COA: cyclooctylamine;
CSA :camphorsulfonic acid;
DCC: *N*,*N*'-dicyclohexylcarbodiimide;
DCE: dynamic contrast-enhanced;
DCM: dichloromethane;
DIDS: 4,4'-diisothiocyanatostilbene-2,2'-disulfonic acid;
DMAP: dimethylaminopyridine;
N,*N*-DMCHA: *N*,*N*-dimethylcyclohexylamine;
DMF: dimethylformamide;
DMSO: dimethylsulfoxide;
doxo: doxorubicin;
eNOS: endothelial nitric oxide synthase;
EC: endothelial cell;
Et: ethyl;
FGF: fibroblast growth factor;

Hb: hemoglobin;
HCC: hepatocellular carcinoma;
HIF: hypoxia-inducible factor;
HPLC: high performance liquid chromatography;
HUVEC: human umbilical vein endothelial cell;
IBPP: inositol bispyrophosphate;
IHC: immunohistochemical;
IHP: inositol hexakisphosphate (IP_6);
ILK: integrin-linked kinase;
iNOS: inducible nitric oxide synthase;
IPK1: $I(3,4,5,6)P_5$ 2-kinase;
IPMK: inositol phosphate multikinase;
ITP: inositol tetrakisphosphate (IP_4);
ITPP: inositol trispyrophosphate;
ITPK1: $I(1,3,4)P_3$ 6-kinase/$I(3,4,5,6)P_4$ 1-kinase;
m-CPBA: *m*-chloroperbenzoic acid;
MDR: multidrug resistance;
Me: methyl;
MEC: microvascular endothelial cell;
miRNA: micro ribonucleic acid;
MMP: matrix metalloproteinase;
MRI: magnetic resonance imaging;
mRNA: messenger ribonucleic acid;
NAP-taurine: *N*-(4-azido-2-nitrophenyl)-2-aminoethylsulfonate;
NMR: nuclear magnetic resonance;
ODC: oxygen dissociation curve;
OEF%: oxygen exchangeable fraction;
P_{50}: partial pressure of O_2 required to achieve 50% saturation of Hb;
PAZ: piperazine;
PCA: prostate cancer;
PI3K: phosphatidylinositol 3-kinase;
PIP_2: phosphatidylinositol-4,5-bisphosphate;
PIP_3: phosphatidylinositol-3,4,5-trisphosphate;
PMA: phorbol-12-myristate 12 acetate;
Ph: phenyl;
Phos: *N*,*N*-diethyl-1,5-dihydro-2,4,3-benzodioxaphosphepin-3-amine;
pO_2: partial oxygen pressure;
Pr: propyl;
PTEN: phosphatase and tensin homologe;

Py: pyridine;
RBC: red blood cell;
r.t.: room temperature;
SAR: structure activity relationship;
SCD: sickle cell disease;
S.D.: standard deviation;
TEA: triethylamine;
TFA: trifluroacetic acid;
TRAMP: transgenic adenocarcinoma of mouse prostate;
VEGF: vascular endothelial growth factor;
VOC: vaso-occlusive crisis;
w: week.

References

[1] Billington, D. C. *The Inositol Phosphates*, Weinheim, Wiley-VCH, 1993.

[2] Potter, B. V. L.; Lampe, D. Chemistry of inositol lipid mediated cellular signaling. *Angew. Chem. Int. Ed.*, 1995, 34, 1933–1972.

[3] Ferguson, M. A. J.; Homans, S. W.; Dwek, R. A.; Rademacher, T. W. Glycosyl-phosphatidylinositol moiety that anchors Trypanosoma brusei variant surface glycoprotein to the membrane. *Science*, 1988, 239, 753–759.

[4] Carmeliet, P. Angiogenesis in health and disease. *Nat. Med.*, 2003, 9, 653–660.

[5] Ferrara, N.; Davis-Smyth, T. The biology of vascular endothelial growth factor. *Endocr. Rev.*, 1997, 18, 4–25.

[6] Folkman, J. Angiogenesis in cancer, vascular, rheumatoid and other disease. *Nat. Med.*, 1995, 1, 27–31.

[7] Folkman, J.; Klagsbrun, U. Angiogenic factors. *Science*, 1987, 235, 442–447.

[8] Folkman, J. What is the evidence that tumors are angiogenesis dependent? *J. Natl. Cancer Inst.*, 1990, 82, 4–6.

[9] Carmeliet, P.; Jain, R. K. Angiogenesis in cancer and other diseases. *Nat. Med.*, 2000, 407, 249–257.

[10] Kerbel, R.; Folkman, J. Clinical translation of angiogenesis inhibitors. *Nat. Rev. Cancer*, 2002, 2, 727–739.

[11] Folkman, J. Clinical applications of research on angiogenesis. *N. Engl. J. Med.*, 1995, 333, 1757–1763.

[12] Griffioen, A. W.; Molema, G. Angiogenesis: potentials for pharmacologic intervention in the treatment of cancer, cardiovascular diseases and chronic inflammation. *Pharmacol. Rev.*, 2000, 52, 237–268.

[13] Folkman, J. How is blood vessel growth regulated in normal and neoplastic tissue? *Cancer Res.*, 1986, 46, 467–473.

[14] Cross, M. J.; Claesson-Welsh, L. FGF and VEGF function in angiogenesis: signalling pathways, biological responses and therapeutic inhibition. *Trends Pharmacol. Sci.*, 2001, 22, 201–207.

[15] O'Reilly, M. S.; Boehm, T.; Shing, Y.; Fukai, N.; Vasios, G.; Lane, W. S.; Flynn, E.; Birkhead, J. R.; Olsen, B. R.; Folkman, J. Endostatin: An endogenous inhibitor of angiogenesis and tumor growth. *Cell*, 1997, 88, 277–285.

[16] Kandel, J.; Bossy-Wetzel, E.; Radvanyi, F.; Klagsbrun, M.; Folkman, J.; Hanahan, D. Neovascularization is associated with a switch to the export of bFGF in the multistep development of fibrosarcoma. *Cell*, 1991, 66, 1095–1104.

[17] Gimenez-Gallego, G.; Cuevas, P. Fibroblast growth factors, proteins with a broad spectrum of biological activities. *Neurol. Res.*, 1994, 16, 313–316.

[18] Cross, M. J.; Claesson-Welsh, L. FGF and VEGF function in angiogenesis: signalling pathways, biological responses and therapeutic inhibition. *Trends Pharmacol. Sci.*, 2001, 22, 201–207.

[19] Ferrara, N. VEGF and the quest for tumour angiogenesis factors. *Nat Rev. Cancer*, 2002, 2, 795–803.

[20] Tan, C.; Cruet-Hennequart, S.; Troussard, A.; Fazli, L.; Costello, P.; Sutton, K.; Wheeler, J.; Gleave, M.; Sanghera, J.; Dedhar, S. Regulation of tumor angiogenesis by integrin-linked kinase (ILK). *Cancer Cell*, 2004, 5, 79–90.

[21] Blancher, C.; Moore, J. W.; Robertson, N.; Harris, A. L. Effects of ras and von Hippel-Lindau (VHL) gene mutations on hypoxia-inducible factor (HIF)-1α, HIF-2α, and vascular endothelial growth factor expression and their regulation by the phosphatidylinositol 3-kinase/ Akt signaling pathway. *Cancer Res.*, 2001, 61, 7349–7355.

[22] Downward, J. Mechanisms and consequences of activation of protein kinase B/Akt. *Curr. Opin. Cell Biol.*, 1998, 10, 262–267.

[23] Shiojima, I.; Walsh, K. Role of Akt signaling in vascular homeostasis and angiogenesis. *Circ. Res.*, 2002, 90, 1243–1250.

[24] Maehama, T.; Dixon, J. E. The tumor suppressor, PTEN/MMAC1, dephosphorylates the lipid second messenger, phosphatidylinositol 3,4,5-trisphosphate. *J. Biol. Chem.*, 1998, 273, 13375–13378.

[25] Milkiewicz, M.; Ispanovic, E.; Doyle, J. L.; Haas, T. L. Regulators of angiogenesis and strategies for their therapeutic manipulation. *Int. J. Biochem. Cell Biol.*, 2006, 38, 333–357.

[26] Teicher, B. A. Hypoxia, tumor endothelium, and targets for therapy. *Adv. Exp. Med. Biol.*, 2005, 566, 31–38.

[27] Pore, N.; Jiang, Z.; Gupta, A.; Cerniglia, G.; Kao, G. D.; Maity, A. EGFR tyrosine kinase inhibitors decrease VEGF expression by both hypoxia-inducible factor (HIF)-1-independent and HIF-1-dependent mechanisms. *Cancer Res.*, 2006, 66, 3197–3204.

[28] Harris, A. L. Hypoxia-a key regulatory factor in tumour growth. *Nature Reviews Cancer*, 2002, 2, 38–47.

[29] Bao, B.; Azmi, A. S.; Ali, S.; Ahmad, A.; Li, Y.; Banerjee, S.; Kong, D.; Sarkar, F. H. The biological kinship of hypoxia with CSC and EMT and their relationship with deregulated expression of miRNAs and tumor aggressiveness. *Biochim. Biophys. Acta*, 2012, 1826, 272–296.

[30] Fu, X. S.; Choi, E.; Bubley, G. J.; Balk, S. P. Identification of hypoxia-inducible factor-1alpha (HIF-1alpha) polymorphism as a mutation in prostate cancer that prevents normoxia-induced degradation. *Prostate*, 2005, 63, 215–221.

[31] Zhong, H.; Chiles, K.; Feldser, D.; Laughner, E.; Hanrahan, C.; Georgescu, M. M.; Simons, J.; Semenza, G. L. Modulation of hypoxia-inducible factor 1a expression by the epidermal growth factor/phosphatidylinositol 3-Kinase/PTEN/AKT/FRAP pathway in human prostate cancer cells: implications for tumor angiogenesis and therapeutics. *Cancer Res.*, 2000, 60, 1541–1545.

[32] Semenza, G. L. Intratumoral hypoxia, radiation resistance, and HIF-1. *Cancer Cell*, 2004, 5, 405–406.

[33] Folkman, J. Angiogenesis: an organizing principle for drug discovery? *Nat. Rev. Drug Discov.*, 2007, 6, 273–286.

[34] Carmeliet, P. Angiogenesis in health and disease. *Nat. Med.*, 2003, 9, 653–660.

[35] Rees, D. C.; Williams, T. N.; Gladwin, M. T. Sickle-cell disease. *Lancet*, 2010, 376, 2018–2031.

[36] Stuart, M. J.; Nagel, R. L. Sickle-cell disease. *Lancet*, 2004, 364, 1343–1360.

[37] Conran, N.; Costa, F. F. Hemoglobin disorders and endothelial cell interactions. *Clin. Biochem.*, 2009, 42, 1824–1838.

[38] Shamsuddin, A. M. Anticancer function of phytic acid. *Int. J. Food Sci. Technol.*, 2002, 37, 769–782.

[39] Shamsuddin, A. M.; Vucenik, I.; Cole, K. E. IP_6: a novel anticancer agent. *Life Sci.*, 1997, 61, 343–354.

[40] Vucenik, I.; Shamsuddin, A. M. Cancer Inhibition by Inositol Hexaphosphate (IP_6) and Inositol: From Laboratory to Clinic. *J. Nutr.*, 2003, 133, 3778S–3784S.

[41] Vucenik, I.; Shamsuddin, A. M. Protection against cancer by dietary IP_6 and inositol. *Nutr. Cancer*, 2006, 55, 109–125.

[42] Shamsuddin, A. M. Metabolism and cellular functions of IP_6: a review. *Anticancer Res.*, 1999, 18, 3733–3736.

[43] Connelly, P. Nutritional advantages and disadvantages of dietary phytates: Part 1. *J. Aust. Tradit. Med. Soc.*, 2011, 17, 21–24.

[44] Posternak, S. Sur la matière phospho-organique de réserve des plantes à chlorophylle. Procédé de préparation. *Comptes Rendus*, 1903, 137, 202–204.

[45] Posternak, M. S. Sur la synthése de l' ether hexaphosphorique de l' inosite et son identité avec la principe phosphor-organique de réserve des plantes vertes. *C. R. Acad. Sci.*, 1919, 169, 138–140.

[46] Posternak, S. Sur la synthèse de l'acide inosito-hexaphosphorique. *Helv. Chim. Acta*, 1921 , 4, 150–165.

[47] Isbrandt, L. R.; Oertel, R. P. Conformational states of myo-inositol hexakis(phosphate) in aqueous solution. A ^{13}C NMR, ^{31}P NMR, and Raman spectroscopic investigation. *J. Am. Chem. Soc.*, 1980, 102, 3144–3148.

[48] Nicolau, C.; Lehn J-M.; Fylaktakidou, K. C.; Greferath, R. Calcium/sodium salt of inositol trispyrophosphate as an allosteric effector of hemoglobin. *Patent* WO 145751 A1, 2009.

[49] Vincent, S. P.; Lehn, J-M.; Lazarte J.; Nicolau, C. Transport of the highly charged *myo*-inositol hexakisphosphate molecule across the red blood cell membrane: A phase transfer and biological study. *Bioorg. Med. Chem.*, 2002, 10, 2825–2834.

[50] Chen, J. L-Carnitine and alkanoyl L-carnitine phytates and process for preparing the same. *Patent* US 0317890 A1, 2010.

[51] Shears, S. B. Diphosphoinositol polyphosphates: Metabolic messengers? *Mol. Pharmacol.*, 2009, 76, 236–252.

[52] Shears, S. B. Molecular basis for the integration of inositol phosphate signaling pathways via human ITPK1. *Advan. Enzyme Regul.*, 2009, 49, 87–96.

[53] Shears, S. B.; Yang, L.; Qian, X. Cell signaling by a physiologically reversible inositol phosphate kinase/phosphatase. *Advan. Enzyme Regul.*, 2004, 44, 265–277.

[54] Shears, S. B. How versatile are inositol phosphate kinases? *Biochem. J.*, 2004, 377, 265–280.

[55] Seeds, A. M.; Sandquist, J. C.; Spana, E. P.; York, J. D. A molecular basis for inositol polyphosphate synthesis in Drosophila melanogaster. *J. Biol. Chem.*, 2004, 279, 4722–4732.

[56] Stevenson-Paulik, J.; Bastidas, R. J.; Chiou, S-T; Frye, R. A.; York, J. D. Generation of phytate-free seeds in Arabidopsis through disruption of inositol polyphosphate kinases. *Proc. Natl. Acad. Sci. USA*, 2005, 102, 12612–12617.

[57] Benesch, R. E.; Edalji, R.; Benesch, R. Reciprocal interaction of hemoglobin with oxygen and protons. The influence of allosteric polyanions. *Biochemistry*, 1977, 16, 2594–2597.

[58] Benesch, R.; Benesch, R. E. Intracellular organic phosphates as regulators of oxygen release by haemoglobin. *Nature*, 1969, 221, 618–622.

[59] Nicolau, C.; Gersonde, K. Incorporation of inositol hexaphosphate into intact red blood cells. *Naturwissenschaften*, 1979, 66, 567–570.

[60] Nicolau, C.; Gersonde, K. Incorporation of inositol hexaphosphate into intact red blood cells - I. Fusion of effector-containing lipid vesicles with erythrocytes. *Naturwissenschaften*, 1979, 66, 563–566.

[61] Gersonde, K.; Nicolau, C. Improvement of the red blood cell O_2 release capacity by lipid vesicle-mediated incorporation in inositol hexaphosphate. *Blut*, 1979, 39, 1–7.

[62] Gersonde, K.; Nicolau, C. Modification of the oxygen affinity of intracellular haemoglobin by incorporation of polyphosphates into intact red blood cells and enhanced O_2 release in the capillary system. *Bibl. Haematol.*, 1980, 46, 81–92.

[63] Kumpati, J.; Franco, R. S.; Weiner, M.; Martelo, O. J. Sickling as a function of oxygen delivery: effect of simulated transfusions of stored, fresh and inositol-hexaphosphate-loaded red cells. *Blood Cells*, 1982, 8, 263–272.

[64] Franco, R. S.; Weinger, M.; Wagner, K.; Martelo, O. J. Incorporation of inositol hexaphosphate into red blood cells mediated by dimethyl sulfoxide. *Life Sci.*, 1983, 32, 2763–2768.

[65] Franco, R. S.; Wagner, K.; Weiner, M.; Martelo, O. J. Preparation of low affinity red cells with dimethylsulfoxide-mediated inositol hexaphosphate incorporation: Hemoglobin and ATP recovery using a continuous-flow method. *Am. J. Hematol.*, 1984, 17, 393–400.

[66] Franco, R. S.; Barker, R.; Novick, S.; Weiner, M.; Martello, O. J. Effect of inositol hexaphosphate on the transient behavior of red cells following a DMSO-induced osmotic pulse. *J. Cell Physiol.*, 1986, 129, 221–229.

[67] Franco, R.; Barker, R.; Mayfield, G.; Silberstein, E.; Weiner, M. The in vivo survival of human red cells with low oxygen affinity prepared by the osmotic pulse method of inositol hexaphosphate incorporation. *Transfusion*, 1990, 30, 196–200.

[68] Teisseire, B. P.; Ropars, C.; Vallez, M. O.; Herigault, R. A.; Nicolau, C. Physiological effects of high-P_{50} erythrocyte transfusion on piglets. *J. Appl. Physiol.,* 1985, 58, 1810–1817.

[69] Ropars, C.; Teisseire, B.; Avenard, G. Improved oxygen delivery to tissues and iron chelator transport through the use of lysed and resealed red blood cells: a new perspective on Cooleys anemia therapy. *Ann. NY Acad. Sci.*, 1985, 445, 304–315.

[70] Teisseire, B.; Ropars, C.; Nicolau, C.; Vallez, M. O.; Chassaigne, M. Enhancement of P_{50} by inositol hexaphosphate entrapped in resealed erythrocytes in piglets. *Adv. Exp. Med. Biol.*, 1984, 180, 673–677.

[71] Ropars, C.; Nicolau, Y.; Chassaigne, M. Apparatus for encapsulating biological active substances into erythrocytes. *Patent* US 4752586 A, 1988.

[72] Ropars, C.; Nicolau, Y. C.; Chassaigne, M. Process and devide to encapsulate in red blood cells at least one biologically active substance, including allosteric effectors of haemoglobin and RBC thus obtained. *Patent* CA 1221653 A2, 1987.

[73] Ropars, C.; Chassaigne, M.; Villereal, M. C.; Avenard, G.; Nicolau, C. Resealed red blood cells as new blood transfusion product. *Bibl. Hematol.,* 1985, 51, 97–105.

[74] Teisseire, B.; Ropars, C.; Villeréal, M. C.; Nicolau, C. Long-term physiological effects of enhanced O_2 release by inositol hexaphosphate-loaded erythrocytes. *Proc. Natl. Acad. Sci. USA*, 1987, 84, 6894–6898.

[75] Ropars, C.; Chassaigne, M.; Avenard, G. Engineered erythrocytes: Influence of P_{50} rightward shift and oxemia on oxygen transport to tissues. *Med. Biol. Eng. Comput.*, 1998, 36, 508–512.
[76] Bailleul, C.; Borrelly-Villereal, M. C.; Chassaigne, M.; Ropars, C. Modification of partial pressure of oxygen (P_{50}) in mammalian red blood cells by incorporation of an allosteric effector of hemoglobin. *Biotechnol. Appl. Biochem.*, 1989, 11, 31–40.
[77] Bourgeaux, V.; Hequet, O.; Campion, Y.; Delcambre, G.; Chevrier, A-M.; Rigal, D.; Godfrin, Y. Inositol hexaphosphate-loaded red blood cells prevent in vitro sickling. *Transfusion*, 2010, 50, 2176–2184.
[78] Lamarre, Y.; Bourgeaux, V.; Pichon, A.; Hardeman, M. R.; Campion, Y.; Hardeman-Zijp, M.; Martin, C.; Richalet, J-P.; Bernaudin, F.; Driss, F.; Godfrin, Y.; Connes, P. Effect of inositol hexaphosphate-loaded red blood cells (RBCs) on the rheology of sickle RBCs. *Transfusion*, 2013, 53, 627–636.
[79] Bourgeaux, V.; Aufradet, E.; Campion, Y.; De Souza, G.; Horand, F.; Bessaad, A.; Chevrier, A-M.; Canet-Soulas, E.; Godfrin, Y.; Martin, C. Efficacy of homologous inositol hexaphosphate-loaded red blood cells in sickle transgenic mice. *Br. J. Haematol.*, 2012, 157, 357–369.
[80] Mouneimne, Y.; Barhoumi, R.; Myers, T.; Slogoff, S.; Nicolau C. Stable rightward shifts of the oxyhemoglobin dissocation curve induced by encapsulation of inositol hexaphosphate in red blood cells using electroporation. *FEBS Lett.*, 1990, 275, 117–120.
[81] Bruggemann, U.; Roux, E. C.; Hannig, J.; Nicolau, C. Low-oxygen-affinity red cells produced in a large-volume, continuous-flow electroporation system. *Transfusion*, 1995, 35, 478–486.
[82] Fox, C. H.; Eberl, M. Phytic acid (IP6), novel broad-spectrum anti-neoplastic agent: a systematic review. *Complement. Ther. Med.*, 2002, 10, 229–234.
[83] Kolappaswamy, K.; Williams, K. A.; Benazzi, C.; Sarli, G.; McLeod Jr., C. G.; Vucenik, I.; DeTolla, L. J. Effect of inositol hexaphosphate on the development of UVB-induced skin tumors in SKH1 hairless mice. *Comp. Med.*, 2009, 59, 147–152.
[84] Singh, R. P.; Agarwal, R. Prostate cancer and inositol hexaphosphate: Efficacy and mechanisms. *Anticancer Res.*, 2005, 25, 2891–2904.
[85] Singh, R. P.; Agarwal, R. Mechanisms of action of novel agents for prostate cancer chemoprevention. *Endocr. Relat. Cancer*, 2006, 13, 751–778.

[86] Bohn, L.; Meyer, A. S.; Rasmussen, S. K. Phytate: impact on environment and human nutrition. A challenge for molecular breeding. *J. Zhejiang Univ. Sci. B*, 2008, 9, 165–191.

[87] Vučenik, I.; Stains, J. Cancer preventive and therapeutic properties of IP_6: Efficacy and mechanisms. *Periodicum Biologorum*, 2010, 112, 451–458.

[88] Matejuk, A.; Shamsuddin, A. IP_6 in cancer therapy: Past, present and future. *Curr. Cancer Ther. Rev.*, 2010, 6, 1–12.

[89] Vucenik, I.; Shamsuddin, A. M. Protection against cancer by dietary IP_6 and inositol. *Nutr. Cancer*, 2006, 55, 109–125.

[90] Vucenik, I.; Passaniti, A.; Vitolo, M. I.; Tantivejkul, K.; Eggleton, P.; Shamsuddin, A. M. Anti-angiogenic activity of inositol hexaphosphate (IP_6). *Carcinogenesis*, 2004, 25, 2115–2123.

[91] McMillan, B.; Riggs, D. R.; Jackson, B. J.; Cunnigham, C.; McFadden, D. W. Dietary influence on pancreatic cancer growth by catechin and inositol hexaphosphate. *J. Surg. Res.*, 2007, 141, 115–119.

[92] Singh, R. P.; Sharma, G.; Mailikarjuna, G. U.; Dhanalakshmi, S.; Agarwal, C.; Agarwal, R. In vivo suppression of hormone-refractory prostate cancer growth by inositol hexaphosphate: Induction of insulin-like growth factor binding protein-3 and inhibition of vascular endothelial growth factor. *Clin. Cancer Res.*, 2004, 10, 244–250.

[93] Gu, M.; Roy, S.; Raina, K.; Agarwal, C.; Agarwal, R. Inositol hexaphosphate suppresses growth and induces apoptosis in prostate carcinoma cells in culture and nude mouse xenograft: PI3K-Akt pathway as potential target. *Cancer Res.*, 2009, 69, 9465–9472.

[94] Raina, K.; Ravichandran, K.; Rajamanickam, S.; Huber, K. M.; Serkova, N. J.; Agarwal, R. Inositol hexaphosphate inhibits tumor growth, vascularity, and metabolism in TRAMP mice: A multiparametric magnetic resonance study. *Cancer Prev. Res.*, 2013, 6, 40–50.

[95] Kandzari, S. J.; Riggs, D.; Jackson, B.; Luchey, A.; Oliver, C.; Zaslau, S. In vitro regulation of cell growth and angiogenesis by inositol hexaphosphate in bladder cancer. *Curr. Urol.*, 2012, 6, 199–204.

[96] Morrison, R. S.; Shi, E.; Kan, M.; Yamaguchi, F.; McKeehan, W.; Nawrot, M. R.; Palczewski, K. Inositolhexakisphosphate ($INSP_6$): an antagonist of fibroblast growth factor receptor binding and activity. *In Vitro Cell Dev. Biol.*, 1994, 30A, 783–789.

[97] Kumar, S. M.; Wang, H-M.; Mohan, S. K.; Chou, R-H.; Yu, C. Molecular level interaction of the human acidic fibroblast growth factor

with the antiangiogenic agent, inositol hexaphosphate. *Biochemistry*, 2010, 49, 10756–10764.

[98] Kapral, M.; Wawszczyk, J.; Hollek, A.; Dymitruk, D.; Węglarz, L. Inhibitory effect of inositol hexaphosphate on metalloproteinases transcription in colon cancer cells stimulated with phorbol-12-myristate 13-acetate. *Acta Polon. Pharm - Drug Res.*, 2012, 69, 1307–1312.

[99] Johnson, L. F.; Tate, M. E. Structure of "phytic acids". *Can. J. Chem.*, 1969, 47, 63–73.

[100] Fylaktakidou, K. C.; Lehn, J-M.; Greferath R.; Nicolau C. Inositol tripyrophosphate: a new membrane permeant allosteric effector of haemoglobin. *Bioorg. Med. Chem. Lett.*, 2005, 15, 1605–1608.

[101] Duarte, C. D.; Greferath, R.; Nicolau, C.; Lehn, J-M. myo-Inositol Trispyrophosphate: A Novel Allosteric Effector of Hemoglobin with High Permeation Selectivity across the Red Blood Cell Plasma Membrane. *ChemBioChem.*, 2010, 11, 2543–2548.

[102] Kieda, C.; Greferath, R.; Crola Da Silva, C.; Fylaktakidou, K. C.; Lehn, J-M.; Nicolau, C. Suppression of hypoxia-induced HIF-1{alpha} and of angiogenesis in endothelial cells by myo-inositol trispyrophosphate-treated erythrocytes. *Proc. Natl. Acad. Sci. USA*, 2006, 103, 15576–15581.

[103] Sihn, G.; Walterb, T.; Klein, J-C.; Queguiner, I.; Iwao, H.; Nicolau, C.; Lehn, J-M.; Corvol, P.; Gasc, J-M. Anti-angiogenic properties of myo-inositol trispyrophosphate *in ovo* and growth reduction of implanted glioma. *FEBS Lett.*, 2007, 581, 962–966.

[104] Kieda, C.; El Hafny-Rahbi, B.; Collet, G.; Lamerant-Fayel, N.; Grillon, C.; Guichard, A.; Dulak, J.; Jozkowicz, A.; Kotlinowski, J.; Fylaktakidou, K. C.; Vidal, A.; Auzeloux, P.; Miot-Noirault, E.; Beloeil, J-C.; Lehn, J-M.; Nicolau, C. Stable tumor vessel normalization with pO_2 increase and endothelial PTEN activation by inositol trispyrophosphate brings novel tumor treatment. *J. Mol. Med.*, 2013, article in press, DOI: 10.1007/s00109-013-0992-6.

[105] Aprahamian, M.; Bour, G.; Akladios, C. Y.; Fylaktakidou, K.; Greferath, R.; Soler, L.; Marescaux, J.; Egly, J-M.; Lehn, J-M.; Nicolau, C. Myo-InositolTrisPyroPhosphate Treatment Leads to HIF-1α Suppression and Eradication of Early Hepatoma Tumors in Rats. *ChemBioChem.*, 2011, 12, 777–783.

[106] Derbal-Wolfrom, L.; Pencreach, E.; Saandi, T.; Aprahamian, M.; Martin, E.; Greferath, R.; Tufa, E.; Choquet, P.; Lehn, J-M.; Nicolau, C.; Duluc, I.; Freund, J-N. Increasing the oxygen load by treatment with

myo-inositol trispyrophosphate reduces growth of colon cancer and modulates the intestine homeobox gene Cdx2. Oncogene, 2012, article in press, DOI: 10.1038/onc.2012.445.

[107] Jain, R. K. Normalization of tumor vasculature: an emerging concept in antiangiogenic therapy. *Science*, 2005, 307, 58–62.

[108] Semenza, G. L. Hypoxia-inducible factors: mediators of cancer progression and targets for cancer therapy. *Trends Pharmacol. Sci.*, 2012, 33, 207–214.

[109] Sato, Y. Persistent vascular normalization as an alternative goal of anti-angiogenic cancer therapy. *Cancer Sci.*, 2011, 102, 1253–1256.

[110] Biolo, A.; Greferath, R.; Siwik, D. A.; Qin, F.; Valsky, E.; Fylaktakidou, K. C.; Pothukanuri, S.; Duarte, C. D.; Schwarz, R. P.; Lehn, J-M.; Nicolau, C.; Colucci, W. S. Enhanced exercise capacity in mice with severe heart failure treated with an allosteric effector of hemoglobin, *myo*-inositol trispyrophosphate. *Proc. Natl. Acad. Sci. USA*, 2009, 106, 1926–1929.

[111] Stephens, L. R.; Hawkins, P. T.; Stanley, A. F.; Moore, T.; Poyner, D. R.; Morris, P. J.; Hanley, M. R.; Kay R. R.; Irvine, R. F. myo-Inositol pentakisphosphates. Structure, biological occurrence and phosphorylation to myo-inositol hexakisphosphate. *Biochem. J.*, 1991, 275, 485–499.

[112] Bartlett, G. R. Isolation and assay of red-cell inositol polyphosphates. *Anal. Biochem.*, 1982, 124, 425–431.

[113] Ozaki, S.; Koga, Y.; Ling, L.; Watanabe, Y.; Kimura, Y.; Hirata, M. Synthesis of 2-substituted *myo*-inositol 1,3,4,5-tetrakis(phosphate) and 1,3,4,5,6-pentakis(phosphate) analogs. *Bull. Chem. Soc. Jpn.*, 1994, 67, 1058–1063.

[114] Sung-Kee, C.; Yuong-Tae, C. Synthesis of all possible regioisomers of *myo*-Inositol pentakisphosphate. *Bioorg. Med. Chem. Lett.*, 1996, 6, 2039–2042.

[115] Podeschwa, M. A. L.; Plettenburg, O.; Altenbach, H-J. Flexible Stereo- and Regioselective Synthesis of *myo*-Inositol Phosphates (Part 1): Via Symmetrical Conduritol B Derivatives. *Eur. J. Org. Chem.*, 2005, 3101–3115.

[116] Godage, H. Y.; Riley, A. M.; Woodman, T. J.; Potter, B. V. L. Regioselective hydrolysis of *myo*-inositol 1,3,5-orthobenzoate via a 1,2-bridged 29-phenyl-19,39-dioxolan-29-ylium ion provides a rapid route

to the anticancer agent Ins(1,3,4,5,6)P_5. *Chem. Commun.*, 2006, 2989–2991.

[117] Maffucci, T.; Piccolo, E.; Cumashi, A.; Iezzi, M.; Riley, A. M.; Saiardi, A.; Godage, H. Y., Rossi, C.; Broggini, M.; Iacobelli, S.; Potter, B. V. L.; Innocenti, P.; Falasca, M. Inhibition of the phosphatidylinositol 3-kinase/Akt pathway by inositol pentakisphosphate results in antiangiogenic and antitumor effects. *Cancer Res.*, 2005, 65, 8339–8349.

[118] Razzini, G.; Berrie, C. P.; Vignati, S.; Broggini, M.; Mascetta, G.; Brancaccio, A.; Falasca, M. Novel functional PI 3-kinase antagonists inhibit cell growth and tumorigenicity in human cancer cell lines. *FASEB J.*, 2000, 14, 1179–1187.

[119] Ferry, S.; Matsuda, M.; Yoshida, H.; Hirata, M. Inositol hexakisphosphate blocks tumor cell growth by activating apoptotic machinery as well as by inhibiting the Akt/NFnB-mediated cell survival pathway. *Carcinogenesis*, 2002, 23, 2031–2041.

[120] Lu, P. J.; Gou, D. M.; Shieh, W. R.; Chen, C. S. Molecular interactions of endogenous D-myo-inositol phosphates with the intracellular D-myo-inositol 1,4,5-trisphosphate recognition site. *Biochemistry*, 1994, 33, 11586–11597.

[121] Mills, S. J.; Safrany, S. T.; Wilcox, R. A.; Nahorski, S.; Potter, B. V. L. Synthesis of myo-inositol 1,2,4,5-tetrakisphosphate, a Ca^{2+}-mobilising tetrakisphosphate with a potency similar to myo-inositol 1,4,5-trisphosphate. *Bioorg. Med. Chem. Lett.*, 1993, 3, 1505–1510.

[122] Burford, N. T.; Nahorski, S. R.; Chung, S-K.; Chang, Y-T.; Wilcox, R. A. Binding and activity of the nine possible regioisomers of myo-inositol tetrakisphosphate at the inositol 1,4,5-trisphosphate receptor. *Cell Calcium*, 1997, 21, 301–310.

[123] Wilcocks, R. A; Challiss, R. A. J.; Liu, C.; Potter, B. V. L.; Nahorski, S. R. Inositol-1,3,4,5-tetrakisphosphate induces calcium mobilization via the inositol-1,4,5-trisphosphate receptor in SH-SY5Y neuroblastoma cells. *Mol. Pharmacol.*, 1993, 44, 810–817.

[124] Zhu, D. M.; Tekle, E.; Huang, E. C. Y.; Chock, P. B. Inositol tetrakisphosphate as a frequency regulator in calcium oscillations in HeLa cells. *J. Biol. Chem.*, 2000, 275, 6063–6066.

[125] Wilcox, R. A.; Challiss, R. A. J.; Baudin, G.; Vasella, A.; Potter, B. V. L.; Nahorski, S. R. Stereoselectivity of Ins(1,3,4,5)P_4 recognition sites: Implications for the mechanism of the Ins(1,3,4,5)P4-induced Ca^{2+} mobilization. *Biochem. J.*, 1993, 294, 191–194.

[126] Hsu, A. L.; Lu, P. J.; Chen, C-S. Regulation of nuclear calcium uptake by inositol phosphates and external calcium. *Biochem. Biophys. Res. Commun.*, 1998, 243, 653–656.
[127] Ivorra, I.; Gigg, R.; Irvine, R. F.; Parker, I. Inositol 1,3,4,6-tetrakisphosphate mobilizes calcium in Xenopus oocytes with high potency. *Biochem. J.*, 1991, 273, 317–332.
[128] Gawler, D. J.; Potter, B. V. L.; Gigg, R.; Nahorski, S. R. Interactions between inositol tris- and tetrakis-phosphates. Effects on intracellular Ca^{2+} mobilization in SH-SY5Y cells *Biochem. J.*, 1991, 276, 163–167.
[129] Morris, A. P.; Gallacher, D. V.; Irvine, R. F.; Petersen, O. H. Synergism of inositol trisphosphate and tetrakisphosphate in activating Ca^{2+}-dependent K^+ channels. *Nature*, 1987, 330, 653–655.
[130] Putney, J. W.; Takemura, H.; Hughes, A. R.; Horstman, D. A.; Thastrup, O. How do inositol phosphates regulate calcium signaling? *FASEB J.*, 1989, 3, 1899–1905.
[131] Ely, J. A.; Hunyady, L.; Baukal, A. J.; Catt, K. J. Inositol 1,3,4,5-tetrakisphosphate stimulates calcium release from bovine adrenal microsomes by a mechanism independent of the inositol 1,4,5-trisphosphate receptor. *Biochem. J.*, 1990, 268, 333–338.
[132] Guse; A. H.; Roth, E.; Emmrich, F. D-myo-Inositol 1,3,4,5-tetrakisphosphate releases Ca^{2+} from crude microsomes and enriched vesicular plasma membranes, but not from intracellular stores of permeabilized T-lymphocytes and monocytes. *Biochem. J.*, 1992, 288, 489–495.
[133] Luckhoff, A.; Clapham, D. E. Inositol 1,3,4,5-tetrakisphosphate activates an endothelial Ca^{2+}-permeable channel. *Nature*, 1992, 355, 356–358.
[134] Hashii, M.; Hirata, M.; Ozaki, S.; Nozawa, Y.; Higashida, H. Ca^{2+}-influx gated by inositol-3,4,5,6-tetrakisphosphate in NIH/3T3 fibroblasts. *Biochem. Biophys. Res. Commun.*, 1994, 200, 1300–1306.
[135] Hashii, M.; Hirata, M.; Ozaki, S.; Nozawa, Y.; Higashida, Y. Ca^{2+} influx evoked by inositol-3,4,5,6-tetrakisphosphate in ras-transformed NIH/3T3 fibroblasts. *FEBS Lett.*, 1994, 340, 276–280.
[136] Irvine, R. F.; Cullen, P. J. Will the real IP_4 receptor please stand up? *Curr. Biol.*, 1993, 3, 540–543.
[137] Smith, P. Ins(1,3,4,5)P_4 promotes sustained activation of the Ca^{2+}-dependent Cl^- current in isolated mouse lacrimal cells. *Biochem. J.*, 1992, 283, 27–30.

[138] Ho, M. W. Y.; Shears, S. B.; Bruzik, K. S.; Duszyk, M.; French, A. S.; Ins(3,4,5,6)P4 specifically inhibits a receptor-mediated Ca^{2+}- dependent Cl^- current in CFPAC-1 cells. *Am. J. Physiol. Cell Physiol.*, 1997, 272, C1160–C1168.

[139] Xie, W.; Kaetzel, M. A., Bruzik, K. S.; Dedman, J. R.; Shears, S. B.; Nelson, D. J. Inositol 3,4,5,6-tetrakisphosphate inhibits the calmodulin-dependent protein kinase II-activated chloride conductance in T84 colonic epithelial cells. *J. Biol. Chem.*, 1996, 271, 14092–14097.

[140] Koumbis, A. E.; Duarte, C. D.; Nicolau, C.; Lehn, J-M. Tetrakisphosphates and bispyrophosphates of myo-inositol derivatives as allosteric effectors of human hemoglobin: Synthesis, molecular recognition, and oxygen release. *ChemMedChem.*, 2011, 6, 169–180.

[141] Watanabe, Y.; Mitani, M.; Morita, T.; Ozaki, S. Highly efficient protection by the tetraisopropyldisiloxane-1,3-diyl group in the synthesis of myo-inositol phosphates as inositol 1,3,4,6- tetrakisphosphate. *Chem. Soc. Chem. Commun.*, 1989, 482–483.

[142] Meek, J. L.; Davidson, F.;. Hobbs, F. W. Jr. Synthesis of inositol phosphates. *J. Am. Chem. Soc.*, 1988, 110, 2317–2318.

[143] Kozikowski, A. P.; Fauq, A. H.; Wilcox, R. A.; Nahorski, S. R. Chemical synthesis and biological evaluation of 1d-1,2,4,5-$InsP_4$ and its 3-fluorinated counterpart 1d-3-F-1,2,4,5-$InsP_4$ - potent 1d-1,4,5-$InsP_3$-like calcium mobilizing analogues. *Bioorg. Med. Chem. Lett.*, 1995, 5, 1295–1300.

[144] Chen, J.; Dormán, G.; Prestwich, G. D. Asymmetric total synthesis of D-myo-inositol 1,2,4,5-tetrakisphosphate and its P-2-(O-aminopropyl) derivative. *J. Org. Chem.*, 1996, 61, 393–397.

[145] Mills, S. J.; Potter, B. V. L. Synthesis of the enantiomers of myo-inositol 1,2,4,5-tetrakisphosphate, a regioisomer of myo-inositol 1,3,4,5-tetrakisphosphate. *J. Chem. Soc. Perkin Trans. 1*, 1997, 1279–1286.

[146] Chung, S-K.; Shin, B-G.; Chang, Y-T.; Suh, B-C.; Kim, K-T. Synthesis of D- and L-myo-inositol 1,2,4,5-tetrakisphosphate and stereoselectivity of the I(1,4,5)P_3 receptor binding. *Bioorg. Med. Chem. Lett.*, 1998, 8, 659–662.

[147] Adelt, S.; Plettenburg, O.; Dallmann, G.; Ritter, F. P.; Shears, S. B.; Altenbach, H-J.; Vogel, G. Regiospecific phosphohydrolases from Dictyostelium as tools for the chemoenzymatic synthesis of the enantiomers D-myo-inositol 1,2,4-trisphosphate and D-myo-inositol 2,3,6-trisphosphate: Non-physiological, potential analogues of

biologically active D-myo-inositol 1,3,4-trisphosphate. *Bioorg. Med. Chem. Lett.*, 2001, 11, 2705–2708.

[148] Chung, S-K.; Kwon, Y-U.; Shin, J-H.; Chang, Y-T.; Lee, C.; Shin, B-G.; Kim, K-C.; Kim, M-J. Divergent syntheses of all possible optically regioisomers of myo inositol tris- and tetrakisphosphates. *J. Org. Chem.*, 2002, 67, 5626–5637.

[149] Mills, S. J.; Backers, K.; Erneux, C.; Potter, B. V. L. Synthesis of D- and L-myo-inositol 1,2,4,6-tetrakisphosphate, regioisomers of myo-inositol 1,3,4,5 tetrakisphosphate: Activity against Ins(1,4,5)P_3 binding proteins. *Org. Biomol. Chem.*, 2003, 1, 3546–3556.

[150] Carless, H. A. J.; Busia, K. Total synthesis of myo-inositol polyphosphates from benzene via conduritol B derivatives. *Tetrahedron Lett.*, 1990, 31, 3449–3452.

[151] Podeschwa, M. A. L.; Plettenburg, O.; Altenbach, H-J. Flexible stereo- and regioselective synthesis of myo-inositol phosphates (part 1): Via symmetrical conduritol B derivatives. *Eur. J. Org. Chem.*, 2005, 3101–3115.

[152] Bohn, L.; Josefsen, L.; Meyer, A. S.; Rasmussen, S. K. Quantitative analysis of phytate globoids isolated from wheat bran and characterization of their sequential dephosphorylation by wheat phytase. *J. Agric. Food Chem.*, 2007, 55, 7547–7552.

[153] Greiner, R.; Larsson Alminger, M.; Carlsson, N. G.; Muzquiz, M.; Burbano, C.; Cuadrado, C.; Pedrosa, M. M.; Goyoaga, C. Pathway of dephosphorylation of *myo*-inositol hexakisphosphate by phytases of legume seeds. *J. Agric. Food Chem.*, 2002, 50, 6865–70.

[154] Nakano, T.; Joh, T.; Narita, K.; Hayakawa, T. The pathway of dephosphorylation of *myo*-inositol hexakisphosphate by phytases from wheat bran of *Triticum aestivum* L. cv. Nourin #61. *Biosci. Biotechnol. Biochem.*, 2000., 64, 995–1003.

[155] Greiner, R.; Carlsson, N-G. Myo-inositol phosphate isomers generated by the action of a phytate-degrading enzyme from Klebsiella terrigena on phytate. *Can. J. Microbiol.*, 2006, 52, 759–768.

[156] Fylaktakidou, K. C.; Duarte, C. D.; Koumbis, A. E.; Nicolau, C.; Lehn, J-M. Polyphosphates and pyrophosphates of hexopyranoses as allosteric effectors of human hemoglobin: Synthesis, molecular recognition, and effect on oxygen release. *ChemMedChem.*, 2011, 6, 153–168.

[157] Fylaktakidou, K. C.; Duarte, C. D.; Jogireddy, R.; Koumbis, A. E.; Nicolau, C.; Lehn, J-M. Polyphosphates of pentopyranoses and

pentofuranoses as allosteric effectors of human hemoglobin: Synthesis, molecular recognition, and oxygen release. *ChemMedChem.*, 2011, 6, 1495–1508.

[158] Blum, C.; Karlsson, S.; Schlewert, G.; Spiess, B.; Rehnberg, N. Synthesis of Optically active (+)-D-3,4,5-tri O-phenylcarbamoyl *myo*-inositol from phytic acid. *Tetrahedron Lett.*, 1995, 36, 7239–7242.

[159] Blum, C.; Rehnberg, N.; Spiess, B.; Schlewer, G. Synthesis of optically active *myo*-inositol derivatives starting from phytic acid. *Carbohydr. Res.*, 1997, 302, 163–168.

[160] Sirén, M. Inositoltriphosphate, a method for preparing same and a composition containing same. *Patent* EP 0179440, 1986.

[161] Andersch P.; Schneider M. P. Enzyme assisted synthesis of *D-myo*-inositol-1,2,6-trisphosphate. *Tetrahedron: Asymmetry,* 1996, 7, 349–352.

[162] Chung, S-K.; Kwon, Y-U.; Shin, J-H.; Chang, Y-T.; Lee, C.; Shin, B-G.; Kim, K-C.; Kim, M-J. Divergent syntheses of all possible optically active regioisomers of *myo*-inositol tris- and tetrakisphosphates. *J. Org. Chem.*, 2002, 67, 5626–5637.

[163] Greiner, R.; Larsson Alminger, M. Stereospecificity of *myo*-inositol hexakisphosphate dephosphorylation by phytate-degrading enzymes of cereals. *J. Food Biochem.*, 2001, 25, 229–248.

[164] Hayakawa, T.; Suzuki, K.; Miura, H.; Ohno, T.; Igaue, I. *myo*-Inositol polyphosphate intermediates in the dephosphorylation of phytic acid by acid phosphatase with phytase activity from rice bran. *Agr. Biol. Chem.*, 1990, 54, 279–286.

[165] Greiner, R.; Larsson Alminger, M.; Carlsson, N. G.; Muzquiz, M.; Burbano, C.; Cuadrado, C.; Pedrosa, M. M.; Goyoaga, C. Pathway of dephosphorylation of *myo*-inositol hexakisphosphate by phytases from legume seeds. *J. Agric. Food Chem.*, 2002, 50, 6865–6870.

[166] Maiti, I. B.; Majumber, A. L.; Biswas, B. B. Purification and mode of action of phytase from *Phaseolus aureus*. *Phytochemistry*, 1974, 13, 1047–1051.

[167] Greiner, R.; Larsson Alminger, M.; Carlsson, N. G. Stereospecificity of *myo*-inositol hexakisphosphate dephosphorylation by a phytate-degrading enzyme of baker′s yeast. *J. Agric. Food Chem.*, 2001, 49, 2228–2233.

[168] Cosgrove, D. J. Inositol phosphate phosphatase of microbiological origin. Inositol pentaphosphate intermediates in the dephosphorylation of the hexaphosphates of *myo*-inositol, *scyllo*-inositol, and D-*chiro*-

inositol, by a bacterial (*Pseudomonas* sp.) phytase. *Austr. J. Biol. Sci.*, 1970, 23, 1207–1220.

[169] Greiner, R.; Farouk, A.; Larsson Alminger, M.; Carlsson, N. G. The pathway of dephosphorylation of *myo*-inositol hexakisphosphate by phytate-degrading enzymes of different *Bacillus* spp. *Can. J. Microbiol.*, 2002, 48, 986–994.

[170] Low, P. A.; Schmelzer, J. D.; Ward K. K.; Yao, J. K. Experimental chronic hypoxic neuropathy: relevance to diabetic neuropathy. *Am. J. Physiol. Endocrinol. Metab.*, 1986, 250, E94–E99.

[171] Carrington, A. L.; Calcutt, N. A.; Ettlinger, C. B.; Gustafsson, T.; Tomlinson, D. R. Effects of treatment with myo-inositol or its 1,2,6-trisphosphate (PP56) on nerve conduction in streptozotocin-diabetes. *Eur. J. Pharmacol.*, 1993, 237, 257–263.

In: Inositol
Editors: H. Rocha and M. Cardoso
ISBN: 978-1-62808-520-4

Chapter 2

Inositol Derivatives in Mycobacterium Tuberculosis: Function, Biosynthesis and Therapeutic Implications

***Xinyi Huang*[1] *and Marcy Hernick*[*1,2]**
[1]Department of Biochemistry, Virginia Tech, Blacksburg, VA [2]Department of Pharmaceutical Sciences, Appalachian College of Pharmacy, Oakwood, VA

Abstract

Inositol (Ins) is a polyol used for the preparation of Ins derivatives that carry out essential functions in eukaryotes. Although rarely found in bacterial species there are Ins derivatives that play critical roles in actinomycetes, a group of Gram-positive bacteria that includes *Mycobacterium* species such as *M. tuberculosis*, the causative agent of tuberculosis (TB). Consequently, there is much interest in understanding the functions and biosynthesis of Ins derivatives in mycobacteria to gain insights that can be used for the development of therapeutic agents. Mycobacteria can either obtain Ins from the environment through Ins transporters or synthesize it *de novo* via a two-step enzymatic pathway. Cellular Ins is converted enzymatically into phosphatidylinositol (PI), a

[*] Corresponding author: email: mhernick@acp.edu.

precursor for more complex glycolipids, or incorporated as L-*myo*-inositol 1-phosphate (Ins-1P) into the small molecule mycothiol (MSH). These Ins derivatives carry out critical functions required for mycobacterial viability and virulence. The mycobacterial cell envelop contains an abundance of mannosyl-phosphatidylinositol containing glycolipids and lipoglycans, including phosphatidylinositol mannosides (PIMs) lipoarabinomannan, (LAM), and lipomannan (LM). Importantly, PIMs, LAM, and LM have been shown to function as immunomodulatory molecules, promote the entry of mycobacteria into phagocytic cells, and regulate phagosome maturation. Mycobacteria also produce the Ins derivative MSH, which is used by mycobacteria as the primary reducing agent and in the detoxification of xenobiotics. Due to these important functions, mycobacterial Ins derivatives are of therapeutic interest for vaccine development and as sources of potential targets for drug development. This chapter reviews mycobacterial uptake of Ins and biosynthesis of Ins derivatives, known functions of Ins derivatives in *M. tuberculosis*, and the progress to date on targeting Ins derivatives for vaccine and drug development.

1. Introduction

Mycobacterium species are actinobacteria, a group of Gram-positive bacteria with a high G+C ratio. [1] Notably, *Mycobacterium tuberculosis* is the causative agent of human tuberculosis (TB), which infects one-third of the world's population with 8.7 million new cases and 1.4 million deaths worldwide each year. [2] Although the TB mortality rate has decreased by 41% since 1990, the global burden remains enormous and TB is the second leading cause of deaths from infectious diseases worldwide. [2] The persistence of less active *M. tuberculosis* and the emergence of extensively drug-resistant (XDR) strains present significant challenges for existing TB therapies. [3] Currently available antibiotics are ineffective against non-growing mycobacteria, and 10% of latent TB carriers develop active disease later in their lives due to immunodeficiencies. [3] Moreover about 4% of new cases and 20% of previously treated TB cases are estimated to have resistance to several first-line and second-line antibiotics. [2] Consequently, new therapeutic alternatives that can circumvent these challenges are needed.

Although Ins derivatives are ubiquitously found in eukaryotes and play an essential role in cellular signal transduction pathways, Ins metabolism in bacteria and archea is limited. [4] Interestingly, mycobacteria are one group of prokaryotes that have Ins metabolism. Mycobacteria utilize Ins in two unique pathways, namely, in the production of PI lipids and in the biosynthesis of MSH. [5, 6] (Figure 1) Since mutations in the biosynthetic pathways of Ins derivatives severely impact the growth and pathogenesis of *M. tuberculosis*, these pathways may represent novel targets for vaccine and drug development. [5-8] This chapter reviews the current understanding of genes and proteins involved in Ins acquisition and biosynthesis in mycobacteria, the structure of Ins derivatives, as well as their functions in the pathogenesis of this organism, and the therapeutic implications of Ins derivatives in vaccine development and as potential targets for drug development for TB treatment.

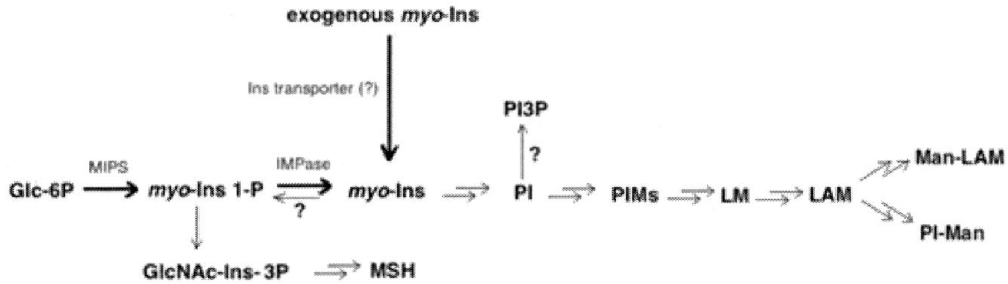

Figure 1. Metabolic pathways for the biosynthesis of inositol derivatives in *Mycobacterium* species. Question marks indicate routes for which the genes encoding the function have not been identified or where the activities of the gene products have not been examined.

2. Inositol Uptake

It has been shown that Ins can be taken up against a concentration gradient in mycobacteria, implicating the presence of an active transport system. [5] A cellular level of 1.2 mM Ins is observed in *Mycobacterium smegmatis* after incubation with radiolabeled *myo*-inositol. [9] The majority of the counts from [^{14}C]-Ins are associated with the pellet fraction that contains the cell wall, while the incorporation of [^{3}H]-Ins is observed in PI and later in PIMs, presumably reflecting incorporation into cell wall components. [9, 10] In the supernatant fraction, [^{14}C]-Ins is rapidly converted to [^{14}C]-Ins-1P and reaches a steady-state level estimated to be 0.2 mM, a value comparable to the apparent K_M (0.15 mM) of MshA, a glycosyltransferase that uses Ins-1P to produce the MSH precursor 1-*O*-(2-acetamido-2-deoxy-α-D-glucopyranosyl)-D-*myo*-inositol-3-phosphate (GlcNAc-Ins-3P) as described below. [9] Cellular

[^{14}C]-MSH levels also increased sharply during incubation, suggesting that the transported *myo*-Ins is converted to Ins-1P and utilized in the production of MSH in *M. smegmatis*. [9].

The *Rv3331* gene in *M. tuberculosis* encodes for a protein that shares 26% identity with the *Bacillus subtitlis* Ins transporter, and therefore, is the most likely candidate for the Ins transporter in *M. tuberculosis*. [11, 12] There are at least six genes in the fast growing species *M. smegmatis* genome that encode for proteins homologous to the Na^+/Ins cotransporter, which are all absent in the *M. tuberculosis* genome. [13] Additionally, two genes downstream of the Ins synthase gene (*ino1*) in *M. smegmatis* that have been suggested to encode for ABC transporters involved in Ins uptake are also missing from the *M. tuberculosis* genome, consistent with the observation that the Ins auxotroph mutant of *M. tuberculosis* requires extremely high concentrations of exogenous Ins supplements (77 mM) for growth. [7] Detailed studies to confirm the *de novo* function of these genes are necessary to understand their roles in TB development.

3. Biosynthesis of Inositol

Ins is synthesized through a two-step pathway that is conserved in all organisms. [4] The enzyme *myo*-inositol 1-phosphate synthase (MIPS) catalyzes the first and rate-limiting step in the pathway - the conversion of D-glucose-6-phosphate (Glc-6P) to Ins-1P (Figure 2), which is also a substrate for MSH biosynthesis. [6] Dephosphorylation of Ins-1P by inositol monophosphatase (IMPase) yields *myo*-Ins (Figure 2), which can further react with CDP-diacyglycerol to generate PI. [14] The virulence of *M. tuberculosis* mutants lacking functional Ins synthetic enzymes is severely attenuated, indicating their essential roles in TB development. Our current understanding of the function and structure of MIPS and IMPase proteins are reviewed below.

3.1. *myo*-inositol 1-Phosphate Synthase

The *Rv0046c* (*Ino1*) gene in *M. tuberculosis* encodes for the MIPS enzyme *myo*-inositol 1-phosphate synthase (tMIPS), which was initially identified as a homolog of MIPS from *Saccharomyces cerevisiae,* or yeast MIPS (yMIPS). [15] (Figure 3) Although the amino acid identity between yMIPS and tMIPS is <15%, *Rv0046c* is able to complement a *S. cerevisiae ino1* mutant, confirming that this gene encodes for tMIPS.[16] The sequence alignment of tMIPS with yMIPS and *Archaeoglobus fulgidus* MIPS (aMIPS) is shown in Figure 3. tMIPS is an essential enzyme for Ins synthesis in *M. tuberculosis*. Inhibition of tMIPS mRNA expression results in enhanced susceptibility of mycobacterial pathogens to antibiotics. [17] Furthermore, *M. tuberculosis* mutants lacking a functional *ino1* can only be isolated with media supplemented with 77 mM Ins, while a *M. smegmatis* mutant lacking a functional *ino1* grows normally in the presence of 0.1 mM Ins. [7] Additionally, it was found that MSH levels fell gradually over a four week period after the *M. tuberculosis ino1* mutant was transferred to an Ins-free medium.[7] Interestingly, Ins taken up from the media must be converted to Ins-1P since Ins cannot directly be used as a substrate for MSH biosynthesis, which suggests the presence of a dedicated inositol kinase or the reverse action of an IMPase; however, no gene encoding for these activities has been identified to date. [5] The virulence of *M. tuberculosis* is severely attenuated in *ino1* mutants, as these mutants are cleared from macrophages, while the CFU of the wild-type strain remains stable. [18] Additionally, the *M. tuberculosis ino1* mutant is incapable of developing disease in a SCID mouse model. [19] The rapid killing of the *ino1* mutant may be attributed to reduced MSH levels, which leave the bacteria susceptible to oxidative stress in macrophages. [7].

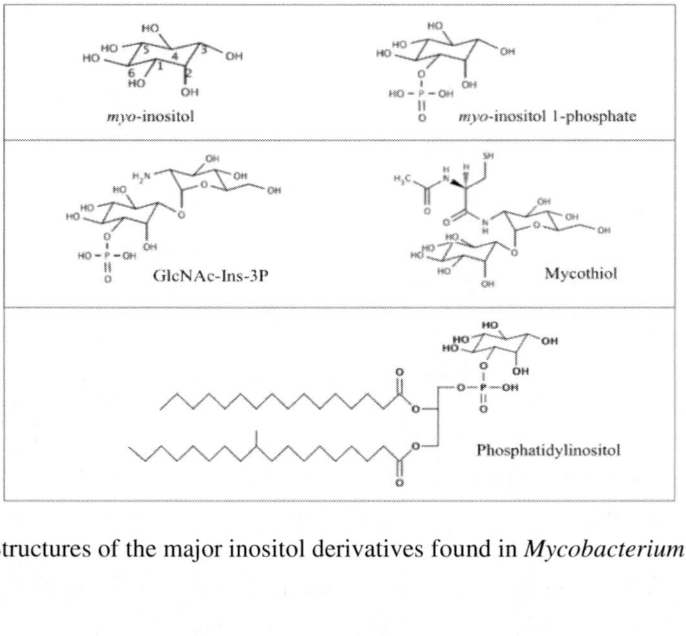

Figure 2. Structures of the major inositol derivatives found in *Mycobacterium* species.

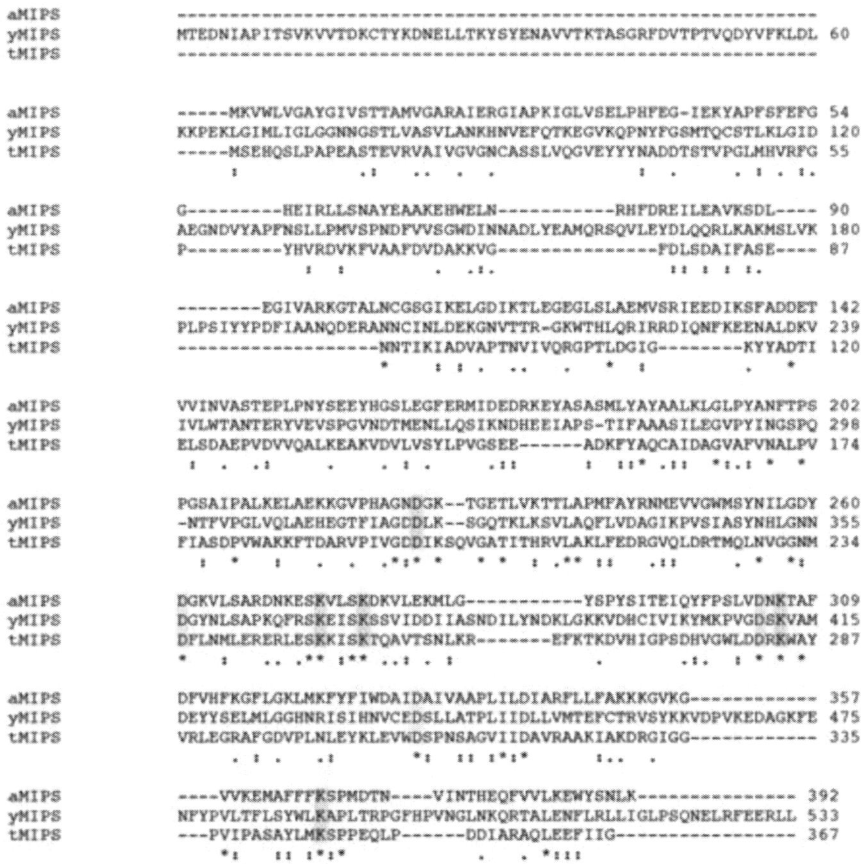

Figure 3. Sequence alignment of MIPS proteins from *A. fulgidus*, *M. tuberculosis,* and *S. cerevisiae* using ClustalW2. '*' identical residues in all sequences; ':' conserved substitutions; '.' semi-conserved substitutions. Conserved lysine residues involved in catalysis are highlighted in green. The second metal ion in aMIPS is proposed to bind to Asp residues that are conserved in all three MIPS and highlighted with yellow.

The crystal structure of tMIPS has been solved as a homotetramer with each unit comprised of two domains (PDB 1GR0). [20] The first domain (D1) is a Rossmann fold and contains the NAD^+ binding site. The second domain (D2) is composed of extensive β-sheets that form the dimer interface (Figure 4) that is structurally similar to aMIPS and much shorter than yMIPS. [20-22] The MIPS active site is lined with highly conserved residues located at the cleft between two domains adjacent to the NAD^+ binding pocket. [20, 21] (Figure 4) A tetrahedral zinc ion is coordinated to the NO7 and NO2 atoms on the NAD^+ nicotinamide moiety, the hydroxyl O atom on S310, and a water molecule that may mimic a hydroxyl group on the substrate in the structure. [20] (Figure 5A) A bound metal ion coordinated to the amide on the nicotinamide is also observed in an analogous location in the crystal structures of yMIPS and aMIPS. [21, 22] (Figure 5B and 5C) This metal ion is likely plays a structural role in positioning the NAD^+ cofactor and is too buried to carry out a catalytic role. [20-23] In addition, only half of the subunits in the tetramer have an occupied metal binding site and NAD^+ exhibits an altered conformation in the absence of bound metal, suggesting negative cooperativity of the tetramer. [21, 22].

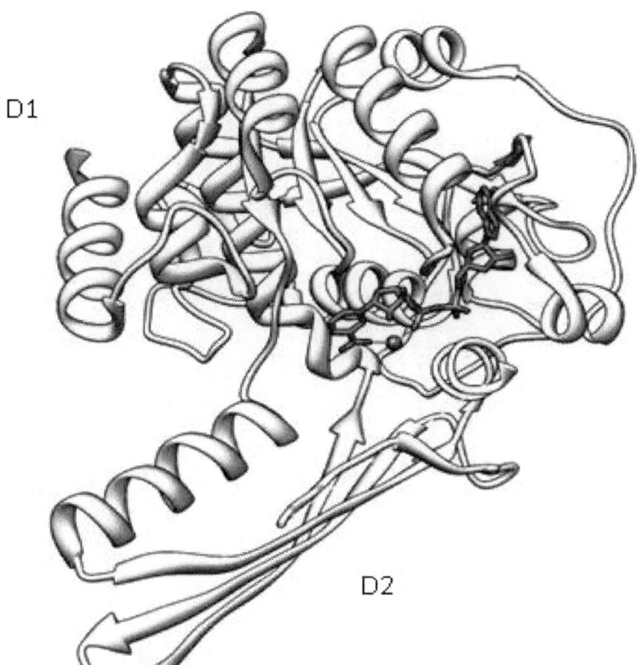

Figure 4. Structure of *M. tuberculosis* MIPS monomer. (PDB 1GR0) The structure contains two domains. D1 contains the NAD^+ (magenta stick) binding site, while D2 forms the dimer interface. A bound zinc ion (green sphere) is also observed at the cleft between two domains.

The *ino1* genes are clustered into two distinct phylogenetic branches, the bacterial and archaeal enzymes (~40 kDa) and the larger (~60 kDa) eukaryotic orthologs. [24-26] Sequence analyses suggest mycobacteria recruited the *ino1* gene from archaea through horizontal gene transfer. [27] While eukaryotic MIPSs including yMIPS are activated by monovalent ammonium ions and unaffected by the presence of EDTA, the aldol cyclization catalyzed by aMIPS follows a type II aldolase mechanism that requires a divalent cation, such as Zn^{2+} or Mg^{2+}, that functions as a Lewis acid and is inhibited by EDTA. [28-30] A second metal ion binding site for the catalytic divalent metal ion has been proposed for aMIPS that occupies a similar position as the putative ammonium ion binding site in yMIPS. [21] This putative divalent metal ion binding site contains five metal ligands including two hydroxyl O atoms (O1 and O2) from the substrate Glc-6P, two carboxylate O atoms from D261 and D332, and one D304 bridged water molecule. [21] Interestingly, the three Asp residues are conserved in tMIPS (D235/D310/D282) and yMIPS (D356/D438/D410) (Figure 3). [21].

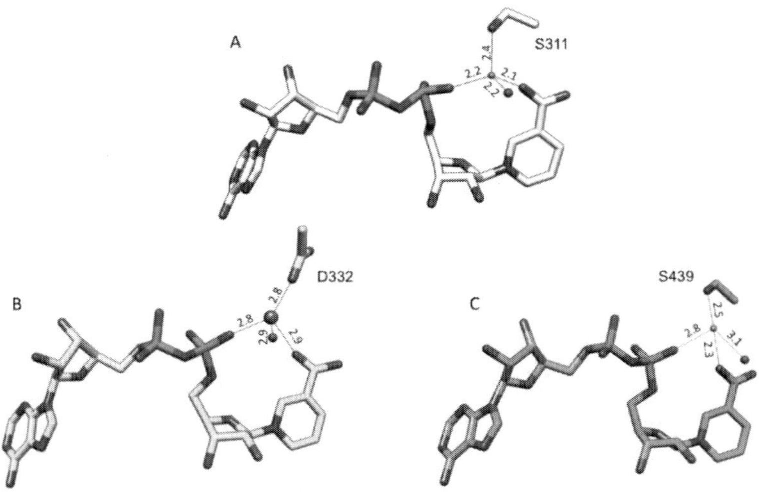

Figure 5. Metal ion binding sites observed in MIPS structures. (A) MIPS from *M. tuberculosis* binds a zinc ion with NAD$^+$, S311 and a water (PDB 1GR0). (B) MIPS from *A. fulgidus* coordinates a potassium ion through NAD$^+$, D332, and a water (PDB 1U1I). (C) MIPS from *S. cerevisiae* contains a manganese ion bound to NAD$^+$, S439, D410, and a water (PDB 1RM0). The interatomic distances (Å) between the metal ions and the ligands are provided.

A chemical mechanism for the tMIPS-catalyzed reaction has been proposed based on sequence homology with aMIPS and yMIPS using the conserved catalytic residues and NAD$^+$ (Figure 6). The acyclic form of the substrate Glc-6P preferentially binds to MIPS [21-23] in an extended conformation with the C5 hydroxyl oriented toward NAD^+, which is held in the active conformation by the bound zinc ion. [22] A proton from the 5-OH of Glc-6P is held by the K248-D197 relay (K274-D225 in aMIPS and K369-D320 in yMIPS) and a hydride is transferred from C5 of Glc-6-P to the reactive C4 of the nicotinamide ring on NAD^+ to generate the keto-Glc-6P intermediate. [21, 22] Subsequent enolization of the C5-C6 bond is catalyzed by K346 (K367 in aMIPS and K489 in yMIPS) that acts as a base to withdraw a *pro*-R proton from C6. [21] Next, a large conformational change of the active site likely occurs bringing C1 in close proximity to C6 allowing for the aldol condensation to occur. [22, 23] The negative charge that accumulates on O1 is stabilized by the conserved K284/K252 (K306/K278 in aMIPS and K369/K373 in yMIPS) and a cation, which is either a divalent metal ion (aMIPS) or an ammonium ion (yMIPS). [21, 28-30] In the last step, direct hydride transfer from NADH reduces C5 in the inosose intermediate to generate Ins-1P. [22] Site-directed mutagenesis of D197, K284, D310 or K346 to Ala results in a complete loss of tMIPS activity [7], which is consistent with the proposed mechanism. It remains to be determined whether tMIPS uses a divalent metal ion or ammonium ion to stabilize the intermediate in aldol condensation.

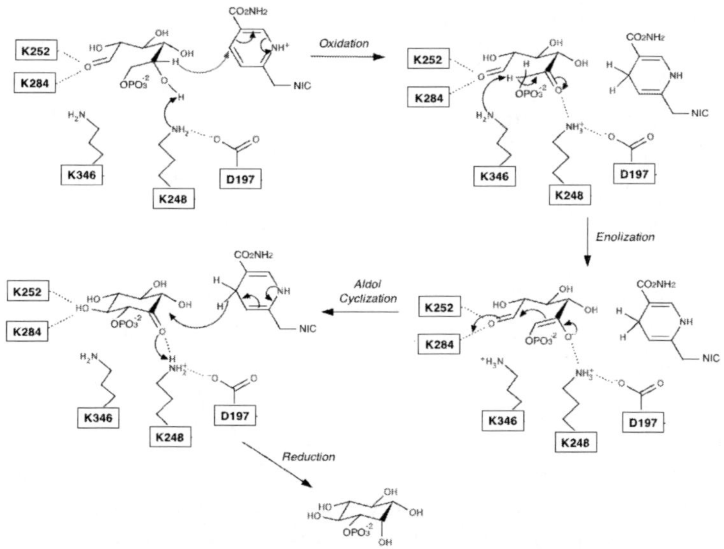

Figure 6. Proposed mechanism for *M. tuberculosis* MIPS based on sequence and structure alignment with *A. fulgidus* and *S. cerevisiae* MIPS.

3.2. Inositol Monophosphatase

There are four inositol monophosphatase (IMPase) homologues in the *M. tuberculosis* genome, namely *impA* (*Rv1604*), *suhB* (*Rv2701c*), *cysQ* (*Rv2131c*), and *impC* (*Rv3137*), which share modest identity to each other (27-32%) and homology (22-30%) with human IMPase. [8, 31, 32] (Figure 7) The mRNA levels of *impA*, *suhB*, *cysQ*, and *impC* are 0.4, 0.11, 0.95, and 0.36 to the control gene *sigA* in *M. tuberculosis*. [8] Sequence alignment indicates a highly conserved active site among all four mycobacterial IMPase homologs, which is composed of two conserved motifs: one located at the N-terminus that contains the residues involved in binding the catalytic metal ions and the phosphate moiety on Ins-1P, and the other near the C-terminus that comprises the rest of the metal binding sites. [8] (Figure7) Mutants in *impA*, *suhB*, or *cysQ* respectively, have been isolated with no significant changes in the overall phosphatase activity of cell lysates or cellular levels of PIMs, LAM, LM and MSH, indicating that these IMPase either have redundant functions or are not involved in Ins biosynthesis in *M. tuberculosis*. [8] Failure to isolate an *impC* mutant in the presence of Ins suggests that this gene may encode for a function unrelated to Ins biosynthesis. [8].

3.2.1. ImpA

The *impA* genes from *M. smegmatis* (*MSMEG_3210*) and *M. tuberculosis* (*Rv1604*) are 70% identical to each other. [32] ImpA has been shown to be critical for cell wall synthesis in *M. smegmatis*, as altered cell envelope morphology and permeability is observed with the *impA* mutant. [32] This is likely a consequence resulting from decreased levels of PI and PIM_2. Without a functional copy of *impA*, only 50% of lipophilic molecules are accumulated in cells compared to the parent strain, consistent with enhanced resistance to hydrophobic antibiotics including chloramphenicol and erythromycin. [32] On the other hand, the function of the *impA* gene remains unclear in *M. tuberculosis*, as the *impA* mutant does not exhibit any differences in cell wall morphology compared to the wild-type strain.[8] One possible explanation for this finding could be the slower growth rate of *M. tuberculosis* which may allow for other IMPases to compensate for ImpA activity in cell wall synthesis. Functional studies on mycobacterial ImpA, especially those examining its substrate specificity, are needed to shed light on its role in growth and virulence of mycobacterial pathogens.

```
Mt_ImpA         --------MHLDSLVAPLVEQASAILDAATALFLVGHRADS-----AVRKKGNDFATEVD 47
Mt_SuhB         MTRPDNEPARLRSVAENLAAEAAAFVRGRRAEVFGISRAGDGDGAVRAKSSPTDPVTVVD 60
Mt_CysQ         -----MVSPAAPDLTDDLTDAELAADLAADAGKLLLQVRAEIG-----FDQPWTLGEAGD 50
Mt_ImpC         ---------------MSHDDLMLALALADRADELTRVRFGALDLRIDTKPDLT-PVTDAD 44
Hum_IMPase      -MKPSGEDQAALAAGPWEECFQAAVQLALRAGQIIRKALTE-EKRVSTKTSAADLVTETD 58
                                       *   .  *  :                .        *

Mt_ImpA         LAIERQVVAALVAAT-GIEVHGEEFGGPAVDSRW-----VWVLDPIDGTINYAA-GSPLA 100
Mt_SuhB         TDTERLLRDRLAQLRPGDPILGEEGGGPADVTATPSDRVTWVLDPIDGTVNFVY-GIPAY 119
Mt_CysQ         RQANSLLLRRLQAERPGDAVLSEEAHDDLARLKSDR---VWIIDPLDGTREFSTPGRDDW 107
Mt_ImpC         RAVESDVRQTLGRDRPGDGVLGEEFGGSTTFTGR-----QWIVDPIDGTKNFVR-GVPVW 98
Hum_IMPase      HLVEDLIISELRERFPSHRFIAEEAAASGAKCVLTHSP-TWIIDPIDGTCNFVH-RFPTV 116
                   :  :   *     .  . .**                *::**:*** ::
                                                   Motif 1

Mt_ImpA         AILLGLLHDGVPVAGLTWMPFTDPRYTAVAG-GPLIKNGVPQPPLADAELANVLVGVGTF 159
Mt_SuhB         AVSIGAQVGGITVAGAVADVAARTVYSAATGLGAHLTDERGR-HVLRCTGVDELSMALLG 178
Mt_CysQ         AVHIALWRR---SSNGQPEITDAAVALPARG-------------NVVYRTDTVTSGAAPA 151
Mt_ImpC         ASLIALLEDGVPSVGVVSAPALQRRWWAARGRGAFASVDGARPHRLSVSSVAELHSASLS 158
Hum_IMPase      AVSIGFAVRQELEFGVIYHCTEERLYTGRRGRGAFCN-----GQRLRVSGETDLSKALVL 171
                *  :.         .               *                         .

Mt_ImpA         SADSRGQFPGRYRLAVLEKLSRVSS---RLRMHGSTGIDLVFVADGILGGAISFG-GHVW 215
Mt_SuhB         TGFG---YSVRCREKQAELLAHVVPLVRDVRRIGSAALDLCMVAAGRLDAYYEHG-VQVW 234
Mt_CysQ         GVPGTLRIAVSATRPPAVLHRIRQTLAIQPVSIGSAGAKAMAVIDGYVDAYLHAGGQWEW 211
Mt_ImpC         FSS----LSGWARPGLRERFIGLTDTVWRVRAYG-DFLSYCLVAEGAVDIAAEPQ-VSVW 212
Hum_IMPase      TEIGPKRDPATLKLFLSNMERLLHAKAHGVRVIGSSTLALCHLASGAADAYYQFG-LHCW 230
                        .                        *        :  *  .          *

Mt_ImpA         DHAAGVALVRAAGGVVTDLAG-QPWTPASRSALAGPPRVHAQILEILGSIGEPEDY-- 270
Mt_SuhB         DCAAGALIAAEAGARVLLSTP--RAGGAGLVVVAAAPGIADELLAALQRFNGLEPIPD 290
Mt_CysQ         DSAAPAGVMLAAGMHASRLDGSPLRYNQLDPYLPDLLMCRAEVAPILLGAIADAWR-- 267
Mt_ImpC         DLAALDIVVREAGGRLTSLDG--VAGPHGGSAVATNGLLHDEVLTRLNAG-------- 260
Hum_IMPase      DLAAATVIIREAGGIVIDTSGGPLDLMACRVVAASTREMAMLIAQALQTINYGRDDEK 288
                * **   :   **                    .        :   *
                    Motif 2
```

Figure 7. Sequence alignment of *M. tuberculosis* IMP homologs with human IMPase using ClustalW2. '*' identical residues in all sequences; ':' conserved substitutions; '.' semi-conserved substitutions. The conserved residues that coordinate the catalytic magnesium ion and that accommodate phosphate moiety on Ins-1P are marked with cyan and yellow, respectively. Conserved residues that mediate the Li^+ inhibition are marked with orange.

3.2.2. SuhB

Mycobacterial *suhB* is named after the *E. coli suhB* gene due to the high sequence similarity shared by these genes (53%). [15] Although the function of *suhB* in *E. coli* remains unclear since there are no detectable Ins derivatives found in this species, recombinant SuhB from *M. tuberculosis* has been biochemically characterized and confirmed to have IMPase activity. [8, 31] However, the *in vivo* function of SuhB in mycobacteria remains unclear and no obvious differences in phenotype are observed for a *suhB* mutant derived from *M. tuberculosis.* [8] SuhB exhibits broad substrate specificity with the highest activity on D-Ins-1-P (activity with L-Ins-1-P has not been tested), followed by inositol-2-phosphate and Glc-6P, which have 40% and 70% the reactivity compared to D-Ins-1-P, respectively. Mannitol-1-phosphate, glycerol-2-phosphate, and 2'-AMP have ~10-20% the reactivity compared to Ins-1P. [31]

SuhB activity is stimulated by divalent metal ions and exhibits a maximum activity in the presence of 6 mM Mg^{2+}, a concentration much higher than its mammalian homologs that require 1mM Mg^{2+}. [31] A conserved site responsible for binding three magnesium ions (Mg^{2+}) has been identified in the crystal structure and is composed of E83, D104, D107, W234 and D235. [33] (Figure 8) Binding of Mg^{2+} to SuhB triggers dimerization of the protein, and the dimerized form of the protein becomes dominant in solution when the concentration of Mg^{2+} is increased from 1 mM to 5 mM. [33] Correlation of SuhB activity with the Mg^{2+} dependence of dimerization suggests that catalytic activity is linked to dimerization of the protein. [33] The conserved active site is situated in a cavity at the N-terminus of the protein that binds Ins-1-P with the Ins moiety stabilized by the side chains of D107 and D235, while the phosphate moiety shares hydrogen bonding interactions with G108 and T109. [33] The activity of SuhB is inhibited by Li^+ with IC_{50} value of 0.9 mM, similar to other IMPases. [31] Structural analysis suggests that the Li^+ competes for one of three catalytic Mg^{2+} sites in the active site. [33] A 10-fold increase in resistance to Li^+ is observed for the L81A mutant [31], indicating that L81 mediates Li^+-sensitivity probably via a hydrophobic contact with the metal ligand W234. This residue is also conserved in ImpC, CysQ, and human IMPase, but is absent in ImpA (Figure 7). [31, 33] The present structure of SuhB may serve as a template for future studies on mycobacterial IMPases.

3.2.3. CysQ

M. tuberculosis CysQ shares 30% identity with its homolog in *E. coli*, and a copy of *Rv2131c* (*cysQ*) can reverse the cysteine auxotroph in an *E. coli cysQ* mutant. [15, 34] CysQ is unlikely to be involved in Ins biosynthesis in mycobacteria since isolation of the *M. tuberculosis cysQ* mutant was unsuccessful even in the presence of 77 mM Ins, a concentration of Ins that is sufficient to support normal growth of the *ino1* mutant. [8] Instead, CysQ has been proposed to function as a 3'-phosphoadenosine-5'-phosphatase (PAPase) in sulfur metabolism *in vivo*. [34] Recombinant CysQ from *M. tuberculosis* catalyzes the dephosphorylation of PAP with a 1,000-fold lower K_M value and 16,000-fold higher k_{cat}/K_M value than those observed with Ins-1P substrate. [34] Furthermore, the *Mtb* gene is able to genetically complement an *E. coli* Δ*cysQ* mutant demonstrating that CysQ is able to dephosphorylate PAP and PAPS *in vivo*. [34] The role of CysQ in sulfur assimilation of *M. tuberculosis* has been suggested by Hatzios, *et al.* [35] In their study, a *cysQ* mutant with a

340 internal base deletion showed decreased levels of Sulfolipid-1, a unique lipid that is exclusively found on the outer envelope of *M. tuberculosis*. Although CysQ may be a potential drug target it is not involved in Ins metabolism and therefore will not be discussed further.

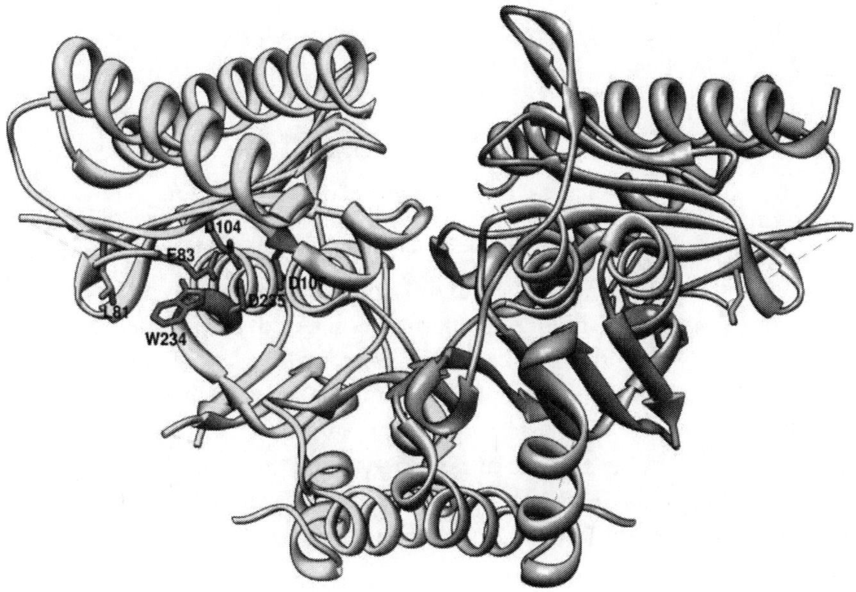

Figure 8. Structure of the *M. tuberculosis* SuhB dimer. The dimer is formed by two identical subunits (yellow and grey) that each contains a conserved metal site that binds three magnesium ions. Side-chains involved in magnesium binding are labeled and shown in magenta sticks. L81 that mediates the Li^+-sensitivity is shown in green stick.

3.2.4. ImpC

The expression level of *impC* increases 3-fold in a *M. tuberculosis ino1* mutant when transferred to Ins-free medium, indicating that *impC* is regulated in response to changes in Ins levels. [8] However, mutation of *impC* is lethal in *M. tuberculosis* even when the media is supplemented with high concentrations of exogenous Ins. [8] The inability to isolate an *impC* mutant is not due to the toxicity of accumulated Ins-1P since an *ino1* mutant background is not sufficient to isolate an *impC* mutant [8] suggesting that the effects of the *impC* mutation are unrelated to Ins production, but are involved in another essential pathway in *M. tuberculosis.* One possible explanation is that the *impC* gene encodes for the MshA2 activity in MSH biosynthesis. [8] MshA2 is an unidentified protein that catalyzes the removal of phosphate from the Ins ring of GlcNAc-Ins-3P to generate 1-*O*-(2-acetamido-2-deoxy-α-D-glucopyranosyl)-D-*myo*-inositol (GlcNAc-Ins). [6] However, since efforts to obtain an *impC* mutant and isolate recombinant ImpC were unsuccessful [8], there is no biochemical evidence regarding the function of ImpC in mycobacteria. Further biochemical and structural studies on the protein encoded by *impC* are necessary to illuminate its physiological function.

4. Inositol Lipids

One distinctive feature of *Mycobacterium* species is the presence of abundant and various Ins-lipid derivatives in the cell wall, which are rarely observed in other bacteria. The structures and functions of mycobacterial Ins-lipid derivatives exhibit unique properties compared to the membrane Ins-lipids extensively found in eukaryotic cells, making them attractive drug targets for TB treatment. Ins-lipid metabolism in mycobacteria has been clarified in detail over the past several years and described in several recent reviews [5, 14, 36]. This chapter summarizes the biosynthesis of Ins-lipids and roles of mycobacterial Ins-lipid glycans in *M. tuberculosis* adaptation in the host, including their interference in macrophage maturation.

4.1. Structure

Ins is the building block for PI (Figure 9), a glycerophospholipid with a glycerol backbone that carries two non-polar fatty acyl residues via carboxylic ester bonds and a molecule of Ins-1P via a phosphoester bond. [4] Importantly, PI from *M. tuberculosis* contains the $C_{19:0}$ fatty acid 10-methyloctadecanoate, which is also known as tuberculosteric acid. [37] In mycobacteria, PI serves as the membrane anchor for glycolipids that play key roles in host-pathogen interactions. [38] PI can be further elaborated to form unique cell wall-associated lipids, including PIMs, LM, and LAM. [39] Results from a recent study suggest that phosphatidylinositol 3-phosphate (PI3P), the membrane-trafficking lipid widely found in eukaryotic cells, is also a downstream product of PI synthesis in *M. smegmatis*. [40].

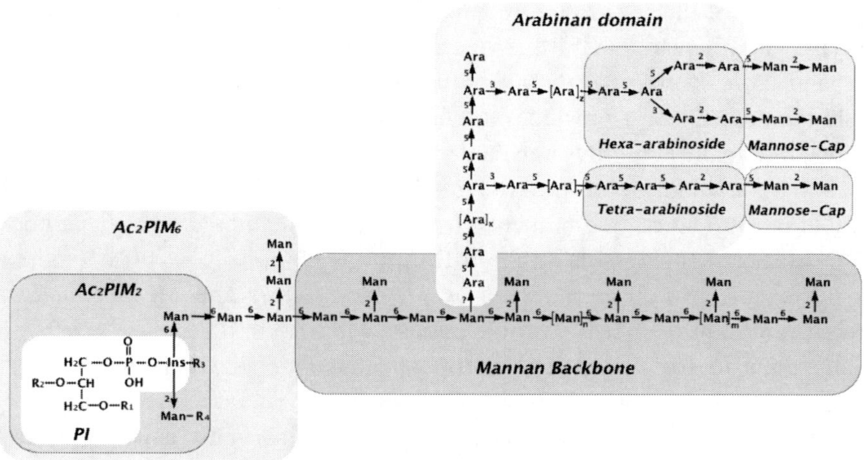

Figure 9. Schematic structures of phosphatidylinositol (PI) and related glycoconjugates. PI is glycosylated at the 2-OH and 6-OH of *myo*-inositol by Man*p* (inset in white), and acylated at 3-OH of inositol and 6-OH of the Man*p* linked at O-2 of inositol in Ac_2PIM_2 (inset in light brown). Man*p* at the 6-OH of inositol is linked to further three and two residues of α(1→6)-Man*p* and α(1→2)-Man*p*, respectively, in Ac_2PIM_6 (inset in cyan). In the mannan backbone of LM, PIM_6 is linked to additional 17–19 residues of α(1→6)-Man*p* and 7–9 singular branched α(1→2)-Man*p* (purple). LM is further linked via an unknown linkage to an arabinan domain containing approximately 70 Ara*f* residues (yellow) branched with α(3→5)-Ara*f* towards its nonreducing end resulting in a linear tetra-arabinoside or/and branched hexa-arabinoside domain (green), which in turn is terminated by β(1→2)-Ara*f* and capped by α(1→2)-Man*p* units (pink). R1, R2, R3 and R4 show different acyl groups in Ac_2PIM_2; n, m, x, y and z represent different degrees of species-specific glycosylation in LM and LAM.

PIMs are mannosylated PI species that contain two or six α-D-mannopyranosyl (Man*p*) moieties attached to the *myo*-Ins ring. [41] (Figure 9) Substitution of the 2-OH and 6-OH of Ins with Man*p* units generates PIM_2, the mannosyl phosphate inositol (MPI) anchor [42]. The heterogeneous nature of MPI results from variations to the nature of fatty acids acylated at 1-OH and 2-OH of the glycerol unit, 3-OH of *myo*-Ins and the 6-OH of the Man*p* residue linked at the O-2 of Ins (see R1, R2, R3 and R4 in Figure 9) [41, 43]. Acylated forms of PIM_2 (Ac_1/Ac_2PIM_2) are both metabolic end products and intermediates in the biosynthesis of high-order PIMs, such as Ac_1/Ac_2PIM_6 [41, 44].

The core structure of LM is composed of 21–34 residues of α (1→6)-Man*p* attached to Ac_1/Ac_2PIM_2 [45] and the mannose backbone is elaborated by 8-9 units of α(1→2)-Man*p* monomer branches, making this oligomannose polymer highly variable [46]. (Figure 9) LAM is built from LM with the mannan core further elaborated by the addition of 55–70 α(1→5)D-arabinofuranose (Ara*f*) residues to the backbone. [46, 47] (Figure 9) The arabinan domain is highly branched with two conserved chain arrangements: linear tetra-arabinoside of the structure β-D-Ara*f* (1→2)-α-D-Ara*f* (1→5)-α-D-Ara*f* (1→5)-α-D-Ara*f* and branched hexa-arabinoside motifs with the structure [β-D-Ara*f* (1→2)-α-D-Ara*f*] 2-3,5-α-D-Ara*f* (1→5)-α-D-Ara*f*. [48] (Figure 9) In pathogenic mycobacteria, such as *M. tuberculosis* and *M. marinum*, the arabinan termini of LAM is further capped with mono, di- and tri-α(1→2)-D-Man*p* units to form Man-LAM [48], and the number of mannose caps is species specific [43, 49]. (Figure 9) This mannose cap structure is not present in avirulent species, which either contain a PI capping motif as identified from *M. smegmatis* (PI-LAM) [49] or are absent of a capping motif as observed in *Mycobacterium chelonae* [50].

4.2. Biosynthesis

4.2.1. PI

In actinobacteria, PI accounts for up to one quarter of total phospholipids [51] and is a bulk membrane phospholipid as well as a precursor for the synthesis of PIMs, LM, and LAM. Constant turnover of PI has been suggested by Haites, *et al*: the [^{3}H]-Ins labeled PI is chased out over 7 hours and a *M. smegmatis* Ins auxotroph depletes the entire pool of PI in 3–5 h. [10] The first step in the production of PI is phosphorylation of diacylglycerol (DAG) by a DAG kinase (*Rv2252*) to form phosphatidic acid [52]. Phosphatidic acid is then activated by CTP through the activity of a CDP-DAG synthase (*Rv2881c*) to form CDP-DAG, a key branch point in lipid biosynthetic pathways [53], which further reacts with *myo*-Ins in a reaction catalyzed by PI synthetase [54]. The *pgsA* gene that encodes PI synthetase is essential for *M. smegmatis* (*MSMEG_2933*) and *M. tuberculosis* (*Rv2612c*) [51], suggesting that PI and/or more complex glycosylated PI are required for normal growth.

4.2.2. PI3P

A recent study has identified the existence of PI3P in *M. smegmatis* [40]. PI3P is an important Ins-lipid in signal transduction that is extensively found in eukaryotic cells, but rarely observed in prokaryotes [55]. The concentration of mycobacterial PI3P is estimated to be ~2.3 μM, corresponding to ~1.6% of total PI in the plasma membrane. Because of its transient production [40], PI3P was initially considered to be an intermediate for PI synthesis through an alternative pathway wherein Ins-1P reacts with CDP-DAG to form PI3P, which is further dephosphorylated to form PI [56]. However, mass spectrometric analysis of *M. smegmatis* crude lysates show that the PI3P contains predominantly $C_{18:1}/C_{19:0}$ fatty acids, in contrast to bulk PI that contains mainly $C_{16:0}/C_{19:0}$ [40]. Therefore, PI3P is not an intermediate for PI synthesis. In addition, inhibition of the dephosphorylation of PI3P results in increased levels of both PI3P and PI [40], which is more consistent with PI3P being a downstream product of PI synthesis.

4.2.3. PIMs

Mycobacterial PIM biosynthesis follows a linear pathway: PI → PIM_2→ PIM_4→ PIM_6. Glycosylation of PI by different α-mannopyranosyltransferases and acetylation by acyltransferases results in the synthesis of Ac_1/Ac_2PIMs [41, 57-63]. The committed step in PIM biosynthesis is catalyzed by PimA (*Rv2610c*), an α-mannopyranosyl-transferase that transfers a mannose from GDP-Man*p* to the 2-OH of the Ins moiety on PI [57]. The downstream *Rv2611c* encodes an acyltransferase (AcylT) that acylates the 6-OH of Man*p* residue linked to the 2-OH of *myo*-Ins in PIM_1 to generate Ac_1PIM_1 [58]. PimB (*Rv2188c*) is an α-D-mannose-α(1→6)-phosphatidylinositol-mannopyranosyltransferase that catalyzes the second mannosylation step to form Ac_1PIM_2. [61-63] PimC is an Ac_1PIM_2:α-D-mannose-α(1→6)-phosphatidylinositol-mannopyranosyl-transferase, which transfers a Man*p* from GDP-Man*p* to the 6-OH of mannose at Ac_1/Ac_2PIM_2, and bioinformatical analyses suggest the existence of redundant genes for PimC [59]. Ac_1/Ac_2PIM_3 is further α(1→6) mannosylated by an unidentified α(1→6)-mannopyranosyltransferase (PimD) to generate Ac_1/Ac_2PIM_4, a key regulatory product involved in Ac_1/Ac_2PIM_6 and/or LM biosynthesis [39, 60, 64]. An α(1→2)-mannopyranosyltransferase PimE (*Rv1159*) utilizes polyprenol-phosphate-mannose (PPM) as a donor to add an α(1→2)-Man*p* to Ac_1/Ac_2PIM_4, resulting in the product Ac_1/Ac_2PIM_5 [60]. It remains unclear whether PimE is solely responsible for the synthesis of both Ac_1/Ac_2PIM_5 and Ac_1/Ac_2PIM_6. In addition, a putative acyl transferase (*Rv1565c*) is suggested in the acylation of higher-order PIMs, LM and LAM, and a *Rv1565c* deletion mutant of *Mycobacterium marinum* shows reduced incorporation of 1,2-[^{14}C]-acetate. [65].

4.2.4. LM

Two α(1→6)-mannopyranosyltransferases use PPM as the mannose donor to extend Ac_1/Ac_2PIM_2 for LM biosynthesis. [64, 66, 67] MptB (*Rv1459c*) is involved in the synthesis of the proximal end of the mannan backbone and speculated to extend an Ac_1/Ac_2PIM_4 acceptor with 12–17 Man*p* units [64], while MptA (*Rv2174*) further elongates the short LM intermediate by synthesizing the distal end of the α(1→6)-Man*p* backbone [66, 67]. The α(1→6)-Man*p* core is further decorated with single α(1→2)-Man*p* branches by a PPM-dependent glycosyltransferases, MptC (*Rv2181*) [14, 46, 68]. In addition, results from a recent study suggest that the elongation of α(1→6)-Man*p* core is controlled by α(1→2) branching via a branching-dependent chain termination mechanism [68].

4.2.5 LAM

Mature LM is considered to be the precursor of LAM by serving as the anchor for arabinan assembly [44]. LM is primed by a few Ara*f* units in a similar fashion as arabinogalactan synthesis in mycobacteria [69], but the enzyme responsible for this activity is unknown. The primed Ara*f*-LM is further extended by EmbC (*Rv3793*) [70, 71] with 12–16 α(1→5)-Ara*f* residues [47]. Several polytopic membrane proteins are also involved in arabinan biosynthesis. AftC (*Rv2673*) introduces the α(1→3)-Ara*f* branch points [47] to generate $[\text{Ara}f]_{12\text{–}16}$-LM. A second branching α(1→3)-arabinofuranosyltransferase, AftD (*Rv0236c*) [72], is likely involved in the synthesis of LAM arabinan. The possibility of AftD functioning as an α(1→5)-arabinofuranosyltransferase involved in α(1→5)-Ara*f* extension of the nonreducing termini of the arabinan domain has also been suggested [72]. The final enzyme in arabinan domain biosynthesis is AftB (*Rv3805c*), a β(1→2)-arabinofuranosyltransferase that generates a terminal tetra- and hexa-arabinofuranoside structure on LAM [47, 73].

4.2.6 Man-LAM

Man-LAM is found in all pathogenic species of *Mycobacterium* and is responsible for some immunomodulatory properties [74]. *Rv1635c* encodes an α(1→5)-mannopyranosyltransferase, CapA, that is involved in Man-LAM capping by addition of the first Man*p* residue to the nonreducing arabinan termini of LAM [75]. The gene ortholog of *Rv1635c* is absent in the *M. smegmatis* genome [5], consistent with its role in Man*p*-capping in pathogenic species. The second, and possibly the third, Man*p* are added by MptC, an α(1→2) mannopyranosyltransferase that is also responsible for branching of LM/LAM [66].

4.3. Function in TB Development

The ability of *M. tuberculosis* to adapt to the host results in latent-TB infections, and ~90% of TB carriers don't develop active disease.[3] Cell wall-associated lipids play key roles in modulating the host immune response during infection and their activity depends on the degree of acylation and mannosylation. [39] The specific interactions between the immune system of the host and each type of Ins-lipid on the mycobacterial cells has been reviewed recently by Mishra, et al. [14]. The presence of the $C_{19:0}$ fatty acid component tuberculosteric acid is a key structural difference that distinguishes *M. tuberculosis* PI to that of other organisms, as it may provide low-level immune response-stimulating potential.[37] Liposomes containing PI and PIMs were shown to activate dendritic cells to produce pro-inflammatory cytokines.[37] LM displays strong proinflammatory effects via Toll-like receptor 2, while LAM inhibits the proinflammatory activity of LM on macrophages.[76, 77] Consequently, the LAM/LM ratio may play a role in establishing bacterial virulence and infection outcomes.[76, 77] This chapter shortly summarizes the role of Ins-lipids in blockade of phagosome maturation in pathogenic *Mycobacterium,* a critical process for their intracellular survival and adaptation to host.

In *M. tuberculosis,* Man-LAM interferes with the phagosome maturation process, by inhibiting fusion of the phagosome with late endosomal and lysosomal organelles, a process that normally leads to the killing and digestion of pathogens in an acidic environment [78]. Normally, the phagosome–lysosome fusion process starts following a cytosolic Ca^{2+} increase. [79] A Ca^{2+}/calmodulin-dependent mechanism sequentially induces the recruitment of a membrane tethering protein early endosome autoantigen 1 (EEA1) to the phagosome [80], and EEA1 induces the delivery of lysosomal hydrolases and H^+-ATPase from the trans-Golgi network to the phagosome. [81] The evasion strategy of blocking phagosome maturation can be mediated by Man-LAM. [82] During mycobacterial uptake, the normal cytosolic Ca^{2+} increase upon an infection is absent [83]. In macrophages infected with *M. tuberculosis*, EEA1 is excluded from the early endosome, thereby inhibiting phagosome maturation. [81] A role for macrophage phosphatase SHP-1 to be activated by Man-LAM and impair Ca^{2+} signaling has been suggested. [79]

Table 1. Genes involved in the biosynthesis of inositol lipids in *Mycobacterium tuberculosis*

ORF	Activity	Function	References
Rv2252	Kinase	Phosphorylates DAG	[52]
Rv2281c	CDP-DAG synthase	Activates DAG by CTP	[53]
Rv2612c (PsgA)	PI synthaase (PI3P synthase?)	Adds *myo*-inositol to CDP-DAG to generate PI	[51]
Rv2610c (PimA)	α(1à2)-Mannopyranosyl-transferase	Adds α(1à2)-Man*p* on inositol moiety of PI to generate PIM_1	[57]
Rv2611c	Acyltransferase	Synthesis of Ac_1/Ac_2PIM_1	[58]
Rv2188c (PimB')	α(1à6)-Mannopyranosyl-transferase	Synthesis of Ac_1/Ac_2PIM_2	[61-63]
PimC	α(1à6)-Mannopyranosyl-transferase	Synthesis of Ac_1/Ac_2PIM_3	[59]
PimD	α(1à6)-Mannopyranosyl-transferase	Synthesis of Ac_1/Ac_2PIM_4	[39, 60, 64]
Rv1565c (PimE)	α(1à2)-Mannopyranosyl-transferase	Synthesis of Ac_1/Ac_2PIM_5 (and Ac_1/Ac_2PIM_6?)	[60]

ORF	Activity	Function	References
Rv1565c	Acyltransferase	Acylates high order PIMs, LM, and LAM	[65]
Rv1459c (MptB)	α(1à6)-Mannopyranosyl-transferase	Synthesis of proximal mannan backbone	[64]
Rv2174 (MptA)	α(1à6)-Mannopyranosyl-transferase	Synthesis of distal mannan backbone	[66, 67]
Rv2181 (MptC)	α(1à2)-Mannopyranosyl-transferase	Adds α(1à2)-Man*p* on the mannan backbone and adds a second mannose cap on ManLAM	[14, 46, 68]
Rv3793 (EmbC)	α(1à5)-arabinofuranosyl-transferase	Synthesis of α(1à5) arabinan backbone	[70, 71]
Rv2673 (AftC)	α(1à3)-arabinofuranosyl-transferase	Adds Ara*f* branch points in α(3à5) direction on arabinan backbone	[47]
Rv0236c (AftD)	α(1à5)-mannopyranosyl-transferase	Adds α(1à5)-Ara*f* branch on the nonreducing termini of arabinan domain	[72]
Rv3805c (AftB)	β(1à2)-arabinofuranosyl-transferase	Generates terminal tetra- and hexa-Araf structure on LAM	[47, 73]

Table 1. (Continued)

ORF	Activity	Function	References
Rv1635c (CapA)	α(1à5)-mannopyranosyl-transferase	Capping of LAM by adding the first Man*p* on the nonreducing arabinan termini	[75]

Man-LAM and related lipoglycans contribute to the MMR-mediated phagocytosis of mycobacteria. [84] MMR (CD206) is a type I transmembrane monomeric C-type lectin with eight carbohydrate recognition domains. [85] MMR signaling may induce an anti-inflammatory response and uptake via the MMR may lead to a latent infection instead of active disease. [84] MMR only recognizes mannose-capped Man-LAM that appears to be restricted to the more virulent *Mycobacterium* species [86], and shows low affinity for avirulent mycobacteria as *M. smegmatis* and *M. phlei* that bear LAM lacking a mannose cap. [87] A *M. marinum* mutant that produces LAM devoid of mannose caps shows a significant increase in co-localization with phagolysosomes in murine macrophages compared to the parent strain. [75] Furthermore, MMR recognizes virulent *M. tuberculosis* strains Erdman and H37Rv, but not the avirulent strain H37Ra, which has subtle structural differences in Man-LAM. [88]

PIMs are also recognized by MMR and play important roles in phagosome maturation via a distinct mechanism from Man-LAM. [89] MMR binds higher-order PIMs and the interactions stimulate fusion of the phagosme with early endosomes by evading acidification. [90] The fusion triggered by PIMs allows mycobacterial pathogens to retrieve nutrients necessary for residing in phagosomal compartments. [90, 91] This indicates a balance between Man-LAM in preventing maturation into the phagolysosome on one hand and PIMs stimulating early endosomal fusion to retrieve nutrients on the other. [91]

The pathway of *de novo* synthesis and the physical functions of PI3P in *Mycobacterium* species are currently unclear. Transient and simulated production of PIP3 in *M. smegmatis* as a response to elevated salt concentrations, but not non-ionic solutes [40], implies that PIP3 possibly serves as a mediator of signaling in mycobacteria similar as observed with eukaryotic cells. It has been shown that *M. tuberculosis* can modulate PI levels and membrane trafficking steps in their mammalian hosts during infection [80, 92], probably through secretion of the mycobacterial enzymes for PI3P synthesis in hosts [5]. Further studies on the presence and synthesis of PI3P in mycobacteria could shed light on the function of PI3P in pathogen-host interaction.

5. Mycothiol

Mycobacterium species produce MSH, a unique low molecular mass thiol that is often present in millimolar amounts, as a surrogate of glutathione

(GSH) found in other organisms. [93] Although it remains controversial as to whether specific genes involved in MSH biosynthesis are essential, it is generally well accepted that the important functions provided by MSH enable *M. tuberculosis* to survive stresses encountered in macrophages. Mutants that produce decreased level of MSH [94] or alternate thiol analogs [95] exhibit enhanced sensitivity to antibiotics and reduced infectivity to macrophages, indicating MSH as a potential target for treatment of latent TB infections. Several reviews [6, 96, 97] have summarized the current progress on the biochemical properties of the MSH enzymes and suggested the potential of MSH synthetic pathways in drug development. This chapter briefly describes the general steps in MSH synthesis, functions of MSH in TB development, and enzymes involved in MSH-dependent pathways.

5.1. Structure and Biosynthesis

MSH contains a cysteine core where the amino group is acetylated and the carboxyl group *N*-linked to D-glucosamine, which is in turn α(1→1) linked to *myo*-inositol (Figure 2), resulting in a unique thiol that is 30-fold and 7-fold more stable than cysteine and GSH, respectively, against heavy metal-catalyzed autoxidation [93]. The biosynthesis of MSH starts with the conversion of Glc-6P to Ins-1P catalyzed by MIPS (*Rv0046c*) [15, 16]. The glycosyltransferase MshA (*Rv0486*) [98, 99] transfers the glucosamine group from UDP-GlcNAc to Ins-1P to form GlcNAc-Ins-3P. Dephosphorylation by a putative phosphatase, MshA2, yields GlcNAc-Ins [9], the substrate for the metal dependent deacetylase MshB (*Rv1170*) [100]. The product of the MshB-catalyzed reaction is 1-*O*-(2-amino-2-deoxy-α-D-glucopyranosyl)-D-*myo*-inositol (GlcN-Ins), the major intermediate found in mycobacteria extracts. Ligation of cysteine with GlcN-Ins by MshC (*Rv2130c*) [101] produces 1-*O*-[2-[[(2R)-2-amino-3-mercapto-1-oxopropyl]amino]-2-deoxy-α-D-glucopyranosyl]-D-*myo*-inositol (Cys-GlcN-Ins). The final step is the acetylation of cysteine catalyzed by MshD (*Rv0819*) to yield MSH. (Figure 10).

The functions of all MSH synthetic enzymes except MshA2 have been genetically confirmed *in vivo* with mutagenesis studies. [7, 98, 102-106] In addition, MIPS, MshB, MshC, and MshD from *M. tuberculosis* have been purified and biochemically examined *in vitro* and the three-dimensional structures for these four enzymes have been solved [20, 100, 101, 107-109]. Although the structure of mycobacterial MshA is not yet available, the structure of MshA from *Corynebacterium glutamicum* (CgMshA), which shares 52% sequence identity to mycobacterial MshA, has been solved and may provide insights about the biochemical properties of mycobacterial MshA. [110] As mentioned above, the gene encoding the MshA2 activity is unidentified.

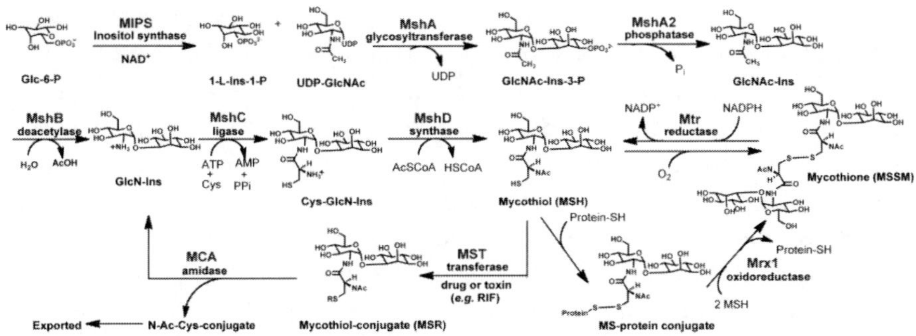

Figure 10. Mycothiol biosynthesis and MSH-dependent detoxification pathways.

5.2. Function in TB Development

MSH is essential for the survival of *M. tuberculosis* in the hostile environment encountered inside macrophage and for adaptation in the host. As the dominant thiol in mycobacterial cytosol, MSH is present in concentrations of ~5 mM in *M. tuberculosis* during the exponential growth phase and 20 mM in the stationary phase. [6] The major function of MSH is to maintain the intracellular redox homeostasis by serving as the cellular thiol buffer with the enzyme mycothione reductase (Mtr, encoded by *Rv2855*) responsible for maintaining MSH in its reduced form.[111] Mycobacterial MSH mutants exhibit enhanced sensitivity to hydrogen peroxide, redox cycling agents as menadione, plumbagin, and gaseous nitric oxide. [104, 112, 113] This ability to protect against damaging effects from reactive oxygen species, a natural by-product of aerobic life, is critical for proper functions of biological processes, including enzyme activation, cell-cycle regulation, and DNA synthesis. [6, 93, 114] MSH also provides protection during adaptation to growth in an oxygen-rich environment in the host, and *M. tuberculosis* mutants with decreased levels of MSH fail to grow in primary murine macrophages. [115, 116]

The constant metabolism of MSH during latent infection is necessary to support mycobacteria survival in the hostile host environment. In *M.*

smegmatis, the half-life for loss of MSH is 50 hours during the stationary phase, indicating the MSH metabolism is constant during the bacterial lifetime. [117] About 10% of the cellular MSH has been found associated with proteins as mycothiol-protein conjugates (MS-protein), which are similar to the glutathione-protein conjugates important for the dormancy of fungal spores. [117, 118]

MSH also plays a role in xenobiotic metabolism. MSH mutants exhibit enhanced sensitivities to antibiotics including rifamycin, streptomycin, erythromycin, and azithromycin, and alkylating agents [104, 113], suggesting that either MSH or MSH-dependent enzymes are involved in protecting mycobacteria from oxidants and toxins. [104] The primary mechanism of MSH-dependent detoxification is the production of less toxic S-conjugates with electrophilic xenobiotics catalyzed by mycothiol S-transferase (MST, encoded by *Rv0443*). [119] The S-conjugates can either accumulate within the cell without adverse consequence [120] or be cleaved by a MS-conjugate amidase (Mca, encoded by *Rv1082*) to generate mercapturic acids (AcCysR) that are exported from the cell, and the resulting GlcN-Ins is recycled to regenerate MSH. [121] Mycoredoxin-1 (Mrx1, encoded by *Rv3198A*) is an oxidoreductase that catalyzes the reduction of S-mycothiolated mixed disulfides, which likely plays a role in the demycothiolation of MSH-protected proteins following oxidative stress.[122]

MSH also functions as the cofactor for several other detoxifying enzymes, including MscR (*Rv2259*), a NAD^+-dependent dehydrogenase that detoxifies either formaldehyde or nitrite through two distinctive mechanisms. [114] Additionally, enhanced sensitivity to peroxides in MSH-deficient *M. smegmatis* and *M. tuberculosis* suggests the presence of an MSH-dependent peroxidase in which MSH acts as an electron donor for peroxiredoxins. [114] MSH has also been associated with the detoxification of methylglyoxal by a glyoxylase I-type activity, and sequence analysis indicates that *Rv0274* encodes a GSH-glyoxylase I homolog. [104].

6. Therapeutic Implications

Mycobacterium species contain a variety of Ins derivatives including PI, the membrane anchor for cell wall lipoglycans (PIMs, LM, and LAM), and MSH, the mycobacterial glutathione analog. [5, 6, 14]. These Ins derivatives play critical roles in the virulence and persistence of *M. tuberculosis*, and therefore, are considered attractive targets for vaccine and drug development.

6.1. Vaccine Development

PIMs, LM, and LAM are important components of mycobacterial cell walls with immunomodulatory activities. Consequently, these molecules, as well as the enzymes that synthesize these molecules, may be useful for the development of TB vaccines. Due to their ability to activate dendritic cells to produce pro-inflammatory cytokines [37] PIMs are potential antigens and/or adjuvants for vaccine development. [37, 123] Furthermore, since the LAM/LM ratio may play a role in establishing bacterial virulence and infection outcomes [76, 77], inactivation of the genes encoding for enzymes that control the LAM/LM ratio may be used to produce attenuated strains of *M. tuberculosis* that can be used for vaccine development. [76].

6.2. Inositol Biosynthetic Pathway

M. tuberculosis mutants lacking functional *ino1* can only be isolated when the media is supplemented with 77 mM Ins, while the equivalent *S. cerevisiae* and *M. smegmatis* mutants only require supplementation with 27 μM and 100 μM Ins, respectively, for normal growth indicating that Ins synthesis or acquisition is essential for the viability of *M. tuberculosis*. [7] Consequently, the Ins biosynthetic pathway may be a source of potential drug targets. Phosphorothioate-modified antisense oligodeoxyribonucleotides (PS-ODNs) against the mRNA of *ino1* are considered potential lipophilic drugs that can pass through the cell wall to enter the cytoplasm of *M. tuberculosis*. Treatment with μM level of antisense PS-ODNs successfully inhibited the expression of tMIPS and the proliferation of *M. tuberculosis*. [17] Detailed characterization of the catalytic process and cofactor preferences of tMIPS is necessary to determine the mechanism utilized by the mycobacterial enzyme, which could aid the rational design of inhibitors against tMIPS. [20, 21] Although four genes homologous to human IMPase were identified only one of these enzymes, ImpC, is considered a potential target for the development of antituberculosis drugs. [8] Future work to identify the biological function of *impC* and characterize the biochemical properties of ImpC are needed for the development of inhibitors and to confirm if ImpC is a viable drug target.

6.3. Inositol-Lipid Pathways

Cell wall components are often the targets of effective chemotherapeutic agents against bacterial infections [124-126]. Ethambutol (EMB), a front-line drug for TB treatment [127], is proposed to specifically target the biosynthesis of arabinan core of LAM [128]. The EMB-resistant mutant derived from *M. smegmatis* makes a truncated LAM, suggesting an effect on arabinosyltransferases. [128] A three-gene operon termed *embCAB* in *M. tuberculosis* has been identified as the putative target of EMB. [129] EmbC is required for synthesis of LAM [70], while *embAB* genes encode Ara*f*T involved in the polymerization of arabinose [130]. Overproduction or structural mutations in Emb protein(s) may cause the high-level resistance to EMB [129]. Furthermore, one EMB derivative, SQ109, has been identified as a novel antibiotic with very efficient antimycobacterial activity [131, 132]. Interestingly, SQ109 shows no cross-resistance to EMB or multidrug resistant strains of *M. tuberculosis*. Other novel and highly potent anti-TB compounds targeting on arabinan synthesis for LAM assembly include benzothiazinones [133] and dinitrobenzamides [134].

Rapid emergence of drug-resistant TB dictates an urgency for the development of more effective drugs that act on novel drug targets with no cross-resistance to existing drugs. The availability of complete genome sequences of *M. tuberculosis* [15] has allowed for the identification of genes and the corresponding enzymes involved in the biosynthesis of Ins-containing glycolipids. The majority of enzymes that play critical roles in the growth and infection of *M. tuberculosis* lack homologs in mammalian systems, thereby making these enzymes attractive drug targets. For example, PI synthetase is critical for the normal growth of *M. smegmatis* and *M. tuberculosis* [51] and therefore is a potential target for inhibitor development. The inability to isolate a *pimA* mutant [57] suggests that mannosylated PI species are also potential targets for drug development against mycobacterial pathogens. Additional sources of targets may be enzymes that are not required for growth, but are critical during TB infection. One example is MptA as mutation of this enzyme results in altered cell wall phenotype with accumulation of a truncated lipoglycan deficient in $\alpha(1\rightarrow6)$Man*p* units [66]. Also, given the importance of the mannose capping structure on Man-LAM in the inhibition of phagosome maturation [75], enzymes such as CapA [75] and MptC [46] are possible targets for the development of antituberculosis agents.

6.4. MSH Biosynthetic and MSH-Dependent Pathways

The MSH biosynthetic and MSH-dependent pathways have become a focal point of drug discovery for TB treatment in recent years. Efforts and progress in development of inhibitors against MSH pathways have been summarized recently by Hernick [96] and are briefly described here. Inhibitors against MshA, MshB, and MshC have already been reported. UDP-(5F)-GlcNAc is a slow, tight-binding inhibitor of MshA, which exhibits competitive inhibition with respect to the substrate UDP-GlcNAc (K_i of ~1.4 μM). [135] Inhibitor libraries against MshB designed to incorporate features of substrate GlcNAc-Ins led to the identification of compounds that inhibit MshB with IC_{50} values low to μM [136, 137]. Intensive efforts have been made to develop MshC inhibitors. 5'-*O*-[*N*-(L-cysteinyl) sulfamoyl]adenosine (CSA) is a bisubstrate analog to MshC that exhibit the inhibition with IC_{50} of ~50 nM [138]. Dequalinium, an ATP-competitive inhibitor that binds to MshC (K_D of ~0.22 μM), inhibits the growth of *M. tuberculosis* under aerobic and anaerobic conditions [139]. NTF1836 (IC_{50} of ~100 μM) is a potent inhibitor against MshC identified in cell-based assays [140]. Treatment with NTF1836 leads to killing nonreplicating *M. tuberculosis* cells and loss of MSH production [140], validating MSH biosynthesis as drug target for latent TB. It should be pointed out that inhibitors of MSH biosynthesis would be precluded from use with antibiotics that require MSH for activity (*e.g.*, INH, ETA).

There have also been efforts to target MSH-dependent enzymes for drug development. Mtr has been targeted for drug development with both subversive substrates (7-methyljuglone derivatives) and inhibitors, with the 7-methyljuglone derivatives possessing potent *in vivo* activity (MIC 0.5 μg/mL). [96] Bromotyrosine containing natural products and analogs inhibit Mca with IC_{50} values in low μM range and show *in vitro* activity against *M. tuberculosis* and *M. bovis*, supporting Mca as a potential drug target. [141-144] In addition, Mca inhibitors designed as substrate analogs have been reported with IC_{50} in the mid μM range [145]. Since MST and Mrx1 were only recently identified, there are no reported inhibitors of these enzymes. Inhibitors targeting enzymes involved in MSH-dependent detoxification may not be precluded from use with antibiotics that require MSH for sensitivity (*e.g.*, INH and ETA), and may offer a means to restore the effectiveness of antibiotics that undergo MSH-dependent detoxification (*e.g.,* RIF and SM). Overall, inhibitors targeting enzymes involved in MSH biosynthesis and MSH-dependent detoxification may be best suited for treatment of latent TB infections either alone or in combination with RIF.

Conclusion

The availability of complete genome sequences of *M. tuberculosis* has allowed for the identification of enzymes and genes involved in the biosynthesis of Ins and related derivatives that are attractive targets for the development of therapeutic agents. Ins-lipids such as PI and PIMs may be useful for the development of TB vaccines either as adjuvants and/or as sources of antigens, while enzymes involved in the biosynthesis of Ins-lipids (*e.g.*, LM, LAM) may be targeted for the production of attenuated *M. tuberculosis* strains for vaccine development. Current challenges in TB therapy remain in the difficulty to eradicate latent and multi-drug resistant pathogens. Exploring novel drug targets involved in TB latency may offer a means to shorten treatment time and decrease the possibility of the emergence of resistant mutant strains. Since Ins derivatives found in *M. tuberculosis* are critical for pathogenesis, playing important roles in mycobacterial persistence and virulence, they are sources of potential drug targets. Enzymes involved in the biosynthesis of Ins-lipid derivatives are possible targets for the treatment of both active and latent TB, while enzymes involved in MSH biosynthesis and MSH-dependent detoxification are best suited as target for the treatment of latent TB. Inhibitors targeting Ins derivatives are attractive since they have the potential to circumvent existing mechanisms of antibiotic resistance.

Abbreviations

aMIPS	*Archaeoglobus fulgidus myo*-inositol 1-phosphate synthase
Ara*f*	α(1→5)D-arabinofuranose
DAG	diacylglycerol
EEA1	early endosome autoantigen 1
EMB	ethambutol
ETA	ethionamide
Glc-6P	D-glucose-6-phosphate
GlcNAc-Ins	1-*O*-(2-acetamido-2-deoxy-α-D-glucopyranosyl)-D-*myo*-inositol
GlcNAc-Ins-3P	1-*O*-(2-acetamido-2-deoxy-α-D-glucopyranosyl)-D-*myo*-inositol-3-phosphate
GSH,	glutathione

IMPase	inositol monophosphatase
INH,	isoniazid
Ins,	inositol
Ins-1P	L-*myo*-inositol 1-phosphate
LAM,	lipoarabinomannan
LM,	lipomannan
Man*p*	α-D-mannopyranosyl
MIPS	*myo*-inositol 1-phosphate synthase
MPI,	mannosyl phosphate inositol
Man-LAM	LAM capped with tri-α(1→2)-D-Man*p* units
Mca,	MS-conjugate amidase
Mrx1,	Mycoredoxin-1
MSH,	mycothiol
MS-protein	mycothiol-protein conjugates
MST,	mycothiol S-transferase
Mtr,	mycothione reductase
PAPase, 3'	phosphoadenosine-5'-phosphatase
PI,	phosphatidylinositol
PI-LAM	LAM with a PI capping motif identified from *M. smegmatis*
PIMs,	phosphatidylinositol mannosides
PI3P,	phosphatidylinositol 3-phosphate
PPM,	polyprenol-phosphate-mannose
PS-ODNs	Phosphorothioate-modified antisense oligodeoxyribonucleotides
TB,	tuberculosis
tMIPS	*Mycobacterium tuberculosis myo*-inositol 1-phosphate synthase
XDR,	extensively drug-resistant
yMIPS	*Saccharomyces cerevisiae myo*-inositol 1-phosphate synthase

References

[1] M. Ventura, C. Canchaya, A. Tauch, G. Chandra, G.F. Fitzgerald, K.F. Chater, and D. van Sinderen, Genomics of Actinobacteria: tracing the evolutionary history of an ancient phylum. *Microbiol Mol Biol Rev* 71 (2007) 495-548.

[2] WHO, Global Tuberculosis Control, 2012.
[3] Y. Zhang, The magic bullets and tuberculosis drug targets. *Annu Rev Pharmacol Toxicol* 45 (2005) 529-64.
[4] R.H. Michell, Inositol derivatives: evolution and functions. *Nat Rev Mol Cell Biol* 9 (2008) 151-61.
[5] Y.S. Morita, T. Fukuda, C.B. Sena, Y. Yamaryo-Botte, M.J. McConville, and T. Kinoshita, Inositol lipid metabolism in mycobacteria: biosynthesis and regulatory mechanisms. *Biochim Biophys Acta* 1810 (2011) 630-41.
[6] G.L. Newton, N. Buchmeier, and R.C. Fahey, Biosynthesis and functions of mycothiol, the unique protective thiol of Actinobacteria. *Microbiol Mol Biol Rev* 72 (2008) 471-94.
[7] F. Movahedzadeh, D.A. Smith, R.A. Norman, P. Dinadayala, J. Murray-Rust, D.G. Russell, S.L. Kendall, S.C. Rison, M.S. McAlister, G.J. Bancroft, N.Q. McDonald, M. Daffe, Y. Av-Gay, and N.G. Stoker, The Mycobacterium tuberculosis ino1 gene is essential for growth and virulence. *Mol Microbiol* 51 (2004) 1003-14.
[8] F. Movahedzadeh, P.R. Wheeler, P. Dinadayala, Y. Av-Gay, T. Parish, M. Daffe, and N.G. Stoker, Inositol monophosphate phosphatase genes of Mycobacterium tuberculosis. *BMC Microbiol* 10 (2010) 50.
[9] G.L. Newton, P. Ta, K.P. Bzymek, and R.C. Fahey, Biochemistry of the initial steps of mycothiol biosynthesis. *J Biol Chem* 281 (2006) 33910-20.
[10] R.E. Haites, Y.S. Morita, M.J. McConville, and H. Billman-Jacobe, Function of phosphatidylinositol in mycobacteria. *J Biol Chem* 280 (2005) 10981-7.
[11] K. Yoshida, Y. Yamamoto, K. Omae, M. Yamamoto, and Y. Fujita, Identification of two myo-inositol transporter genes of Bacillus subtilis. *J Bacteriol* 184 (2002) 983-91.
[12] K.E. Bettaney, P. Sukumar, R. Hussain, G. Siligardi, P.J. Henderson, and S.G. Patching, A systematic approach to the amplified expression, functional characterization and purification of inositol transporters from Bacillus subtilis. *Mol Membr Biol* (2012).
[13] F. Titgemeyer, J. Amon, S. Parche, M. Mahfoud, J. Bail, M. Schlicht, N. Rehm, D. Hillmann, J. Stephan, B. Walter, A. Burkovski, and M. Niederweis, A genomic view of sugar transport in Mycobacterium smegmatis and Mycobacterium tuberculosis. *J Bacteriol* 189 (2007) 5903-15.

[14] A.K. Mishra, N.N. Driessen, B.J. Appelmelk, and G.S. Besra, Lipoarabinomannan and related glycoconjugates: structure, biogenesis and role in Mycobacterium tuberculosis physiology and host-pathogen interaction. *FEMS Microbiol Rev* 35 (2011) 1126-57.

[15] S.T. Cole, R. Brosch, J. Parkhill, T. Garnier, C. Churcher, D. Harris, S.V. Gordon, K. Eiglmeier, S. Gas, C.E. Barry, 3rd, F. Tekaia, K. Badcock, D. Basham, D. Brown, T. Chillingworth, R. Connor, R. Davies, K. Devlin, T. Feltwell, S. Gentles, N. Hamlin, S. Holroyd, T. Hornsby, K. Jagels, A. Krogh, J. McLean, S. Moule, L. Murphy, K. Oliver, J. Osborne, M.A. Quail, M.A. Rajandream, J. Rogers, S. Rutter, K. Seeger, J. Skelton, R. Squares, S. Squares, J.E. Sulston, K. Taylor, S. Whitehead, and B.G. Barrell, Deciphering the biology of Mycobacterium tuberculosis from the complete genome sequence. *Nature* 393 (1998) 537-44.

[16] N. Bachhawat, and S.C. Mande, Identification of the INO1 gene of Mycobacterium tuberculosis H37Rv reveals a novel class of inositol-1-phosphate synthase enzyme. *J Mol Biol* 291 (1999) 531-6.

[17] Y. Li, Z. Chen, X. Li, H. Zhang, Q. Huang, Y. Zhang, and S. Xu, Inositol-1-phosphate synthetase mRNA as a new target for antisense inhibition of Mycobacterium tuberculosis. *J Biotechnol* 128 (2007) 726-34.

[18] D. Smith, H. Hansch, G. Bancroft, and S. Ehlers, T-cell-independent granuloma formation in response to Mycobacterium avium: role of tumour necrosis factor-alpha and interferon-gamma. *Immunology* 92 (1997) 413-21.

[19] H.C. Hansch, D.A. Smith, M.E. Mielke, H. Hahn, G.J. Bancroft, and S. Ehlers, Mechanisms of granuloma formation in murine Mycobacterium avium infection: the contribution of CD4+ T cells. *Int Immunol* 8 (1996) 1299-310.

[20] R.A. Norman, M.S. McAlister, J. Murray-Rust, F. Movahedzadeh, N.G. Stoker, and N.Q. McDonald, Crystal structure of inositol 1-phosphate synthase from Mycobacterium tuberculosis, a key enzyme in phosphatidylinositol synthesis. *Structure* 10 (2002) 393-402.

[21] K.A. Stieglitz, H. Yang, M.F. Roberts, and B. Stec, Reaching for mechanistic consensus across life kingdoms: structure and insights into catalysis of the myo-inositol-1-phosphate synthase (mIPS) from Archaeoglobus fulgidus. *Biochemistry* 44 (2005) 213-24.

[22] X. Jin, and J.H. Geiger, Structures of NAD(+)- and NADH-bound 1-l-myo-inositol 1-phosphate synthase. *Acta Crystallogr D Biol Crystallogr* 59 (2003) 1154-64.

[23] X. Jin, K.M. Foley, and J.H. Geiger, The structure of the 1L-myo-inositol-1-phosphate synthase-NAD+-2-deoxy-D-glucitol 6-(E)-vinylhomophosphonate complex demands a revision of the enzyme mechanism. *J Biol Chem* 279 (2004) 13889-95.

[24] N. Bachhawat, and S.C. Mande, Complex evolution of the inositol-1-phosphate synthase gene among archaea and eubacteria. *Trends Genet* 16 (2000) 111-3.

[25] K.G. Dastidar, A. Chatterjee, A. Chatterjee, and A.L. Majumder, Evolutionary Divergence of L-myo-Inositol 1-Phosphate Synthase: Significance of a "Core Catalytic Structure". in: A.L. Majumder, and B.B. Biswas, (Eds.), Biology of Inositols and Phosphoinositides, Springer US, 2006, pp. 315-340.

[26] R.H. Michell, Inositol and its derivatives: Their evolution and functions. *Advances in Enzyme Regulation* 51 (2011) 84-90.

[27] A.L. Majumder, A. Chatterjee, K. Ghosh Dastidar, and M. Majee, Diversification and evolution of L-myo-inositol 1-phosphate synthase. *FEBS Lett* 553 (2003) 3-10.

[28] L. Chen, C. Zhou, H. Yang, and M.F. Roberts, Inositol-1-phosphate synthase from Archaeoglobus fulgidus is a class II aldolase. *Biochemistry* 39 (2000) 12415-23.

[29] J.E. Escamilla, M. Contreras, A. Martinez, and M. Zentella-Pina, L-myoinositol-1-phosphate synthase from Neurospora crassa: purification to homogeneity and partial characterization. *Arch Biochem Biophys* 218 (1982) 275-85.

[30] K. Neelon, Y. Wang, B. Stec, and M.F. Roberts, Probing the mechanism of the Archaeoglobus fulgidus inositol-1-phosphate synthase. *J Biol Chem* 280 (2005) 11475-82.

[31] J. Nigou, L.G. Dover, and G.S. Besra, Purification and biochemical characterization of Mycobacterium tuberculosis SuhB, an inositol monophosphatase involved in inositol biosynthesis. *Biochemistry* 41 (2002) 4392-8.

[32] T. Parish, J. Liu, H. Nikaido, and N.G. Stoker, A Mycobacterium smegmatis mutant with a defective inositol monophosphate phosphatase gene homolog has altered cell envelope permeability. *J Bacteriol* 179 (1997) 7827-33.

[33] A.K. Brown, G. Meng, H. Ghadbane, D.J. Scott, L.G. Dover, J. Nigou, G.S. Besra, and K. Futterer, Dimerization of inositol monophosphatase Mycobacterium tuberculosis SuhB is not constitutive, but induced by binding of the activator Mg2+. *BMC Struct Biol* 7 (2007) 55.

[34] S.K. Hatzios, A.T. Iavarone, and C.R. Bertozzi, Rv2131c from Mycobacterium tuberculosis is a CysQ 3'-phosphoadenosine-5'-phosphatase. *Biochemistry* 47 (2008) 5823-31.

[35] S.K. Hatzios, M.W. Schelle, G.L. Newton, K.M. Sogi, C.M. Holsclaw, R.C. Fahey, and C.R. Bertozzi, The Mycobacterium tuberculosis CysQ phosphatase modulates the biosynthesis of sulfated glycolipids and bacterial growth. *Bioorg Med Chem Lett* 21 (2011) 4956-9.

[36] M.E. Guerin, J. Kordulakova, P.M. Alzari, P.J. Brennan, and M. Jackson, Molecular basis of phosphatidyl-myo-inositol mannoside biosynthesis and regulation in mycobacteria. *J Biol Chem* 285 (2010) 33577-83.

[37] G.D. Sprott, C.J. Dicaire, K. Gurnani, S. Sad, and L. Krishnan, Activation of Dendritic Cells by Liposomes Prepared from Phosphatidylinositol Mannosides from Mycobacterium bovis Bacillus Calmette-Gu√©rin and Adjuvant Activity In Vivo. *Infection and Immunity* 72 (2004) 5235-5246.

[38] J.B. Torrelles, and L.S. Schlesinger, Diversity in Mycobacterium tuberculosis mannosylated cell wall determinants impacts adaptation to the host. *Tuberculosis* (Edinb) 90 (2010) 84-93.

[39] Y.S. Morita, J.H. Patterson, H. Billman-Jacobe, and M.J. McConville, Biosynthesis of mycobacterial phosphatidylinositol mannosides. *Biochem J* 378 (2004) 589-97.

[40] Y.S. Morita, Y. Yamaryo-Botte, K. Miyanagi, J.M. Callaghan, J.H. Patterson, P.K. Crellin, R.L. Coppel, H. Billman-Jacobe, T. Kinoshita, and M.J. McConville, Stress-induced synthesis of phosphatidylinositol 3-phosphate in mycobacteria. *J Biol Chem* 285 (2010) 16643-50.

[41] K.H. Khoo, A. Dell, H.R. Morris, P.J. Brennan, and D. Chatterjee, Structural definition of acylated phosphatidylinositol mannosides from Mycobacterium tuberculosis: definition of a common anchor for lipomannan and lipoarabinomannan. *Glycobiology* 5 (1995) 117-27.

[42] W.B. Severn, R.H. Furneaux, R. Falshaw, and P.H. Atkinson, Chemical and spectroscopic characterisation of the phosphatidylinositol manno-oligosaccharides from Mycobacterium bovis AN5 and WAg201 and Mycobacterium smegmatis mc2 155. *Carbohydr Res* 308 (1998) 397-408.

[43] J. Nigou, M. Gilleron, and G. Puzo, Lipoarabinomannans: from structure to biosynthesis. *Biochimie* 85 (2003) 153-66.

[44] G.S. Besra, C.B. Morehouse, C.M. Rittner, C.J. Waechter, and P.J. Brennan, Biosynthesis of mycobacterial lipoarabinomannan. *J Biol Chem* 272 (1997) 18460-6.

[45] S.W. Hunter, and P.J. Brennan, Evidence for the presence of a phosphatidylinositol anchor on the lipoarabinomannan and lipomannan of Mycobacterium tuberculosis. *J Biol Chem* 265 (1990) 9272-9.

[46] D. Kaur, A. Obregon-Henao, H. Pham, D. Chatterjee, P.J. Brennan, and M. Jackson, Lipoarabinomannan of Mycobacterium: mannose capping by a multifunctional terminal mannosyltransferase. *Proc Natl Acad Sci U S A* 105 (2008) 17973-7.

[47] H.L. Birch, L.J. Alderwick, B.J. Appelmelk, J. Maaskant, A. Bhatt, A. Singh, J. Nigou, L. Eggeling, J. Geurtsen, and G.S. Besra, A truncated lipoglycan from mycobacteria with altered immunological properties. *Proc Natl Acad Sci U S A* 107 (2010) 2634-9.

[48] D. Chatterjee, K.H. Khoo, M.R. McNeil, A. Dell, H.R. Morris, and P.J. Brennan, Structural definition of the non-reducing termini of mannose-capped LAM from Mycobacterium tuberculosis through selective enzymatic degradation and fast atom bombardment-mass spectrometry. *Glycobiology* 3 (1993) 497-506.

[49] K.H. Khoo, A. Dell, H.R. Morris, P.J. Brennan, and D. Chatterjee, Inositol phosphate capping of the nonreducing termini of lipoarabinomannan from rapidly growing strains of Mycobacterium. *J Biol Chem* 270 (1995) 12380-9.

[50] Y. Guerardel, E. Maes, V. Briken, F. Chirat, Y. Leroy, C. Locht, G. Strecker, and L. Kremer, Lipomannan and lipoarabinomannan from a clinical isolate of Mycobacterium kansasii: novel structural features and apoptosis-inducing properties. *J Biol Chem* 278 (2003) 36637-51.

[51] M. Jackson, D.C. Crick, and P.J. Brennan, Phosphatidylinositol is an essential phospholipid of mycobacteria. *J Biol Chem* 275 (2000) 30092-9.

[52] R.M. Owens, F.F. Hsu, B.C. VanderVen, G.E. Purdy, E. Hesteande, P. Giannakas, J.C. Sacchettini, J.D. McKinney, P.J. Hill, J.T. Belisle, B.A. Butcher, K. Pethe, and D.G. Russell, M. tuberculosis Rv2252 encodes a diacylglycerol kinase involved in the biosynthesis of phosphatidylinositol mannosides (PIMs). *Mol Microbiol* 60 (2006) 1152-63.

[53] J. Nigou, and G.S. Besra, Cytidine diphosphate-diacylglycerol synthesis in Mycobacterium smegmatis. *Biochem J* 367 (2002) 157-62.
[54] M. Salman, J.T. Lonsdale, G.S. Besra, and P.J. Brennan, Phosphatidylinositol synthesis in mycobacteria. *Biochim Biophys Acta* 1436 (1999) 437-50.
[55] S. Cockcroft, Mammalian phosphatidylinositol transfer proteins: emerging roles in signal transduction and vesicular traffic. *Chem Phys Lipids* 98 (1999) 23-33.
[56] H. Morii, M. Ogawa, K. Fukuda, H. Taniguchi, and Y. Koga, A revised biosynthetic pathway for phosphatidylinositol in Mycobacteria. *J Biochem* 148 (2010) 593-602.
[57] J. Kordulakova, M. Gilleron, K. Mikusova, G. Puzo, P.J. Brennan, B. Gicquel, and M. Jackson, Definition of the first mannosylation step in phosphatidylinositol mannoside synthesis. PimA is essential for growth of mycobacteria. *J Biol Chem* 277 (2002) 31335-44.
[58] J. Kordulakova, M. Gilleron, G. Puzo, P.J. Brennan, B. Gicquel, K. Mikusova, and M. Jackson, Identification of the required acyltransferase step in the biosynthesis of the phosphatidylinositol mannosides of mycobacterium species. *J Biol Chem* 278 (2003) 36285-95.
[59] L. Kremer, S.S. Gurcha, P. Bifani, P.G. Hitchen, A. Baulard, H.R. Morris, A. Dell, P.J. Brennan, and G.S. Besra, Characterization of a putative alpha-mannosyltransferase involved in phosphatidylinositol trimannoside biosynthesis in Mycobacterium tuberculosis. *Biochem J* 363 (2002) 437-47.
[60] Y.S. Morita, C.B. Sena, R.F. Waller, K. Kurokawa, M.F. Sernee, F. Nakatani, R.E. Haites, H. Billman-Jacobe, M.J. McConville, Y. Maeda, and T. Kinoshita, PimE is a polyprenol-phosphate-mannose-dependent mannosyltransferase that transfers the fifth mannose of phosphatidylinositol mannoside in mycobacteria. *J Biol Chem* 281 (2006) 25143-55.
[61] D.J. Lea-Smith, K.L. Martin, J.S. Pyke, D. Tull, M.J. McConville, R.L. Coppel, and P.K. Crellin, Analysis of a new mannosyltransferase required for the synthesis of phosphatidylinositol mannosides and lipoarbinomannan reveals two lipomannan pools in corynebacterineae. *J Biol Chem* 283 (2008) 6773-82.
[62] M.E. Guerin, D. Kaur, B.S. Somashekar, S. Gibbs, P. Gest, D. Chatterjee, P.J. Brennan, and M. Jackson, New insights into the early steps of phosphatidylinositol mannoside biosynthesis in mycobacteria:

PimB' is an essential enzyme of Mycobacterium smegmatis. *J Biol Chem* 284 (2009) 25687-96.

[63] A.K. Mishra, S. Batt, K. Krumbach, L. Eggeling, and G.S. Besra, Characterization of the Corynebacterium glutamicum deltapimB' deltamgtA double deletion mutant and the role of Mycobacterium tuberculosis orthologues Rv2188c and Rv0557 in glycolipid biosynthesis. *J Bacteriol* 191 (2009) 4465-72.

[64] A.K. Mishra, L.J. Alderwick, D. Rittmann, C. Wang, A. Bhatt, W.R. Jacobs, Jr., K. Takayama, L. Eggeling, and G.S. Besra, Identification of a novel alpha(1-->6) mannopyranosyltransferase MptB from Corynebacterium glutamicum by deletion of a conserved gene, NCgl1505, affords a lipomannan- and lipoarabinomannan-deficient mutant. *Mol Microbiol* 68 (2008) 1595-613.

[65] N.N. Driessen, E.J. Stoop, R. Ummels, S.S. Gurcha, A.K. Mishra, G. Larrouy-Maumus, J. Nigou, M. Gilleron, G. Puzo, J.J. Maaskant, M. Sparrius, G.S. Besra, W. Bitter, C.M. Vandenbroucke-Grauls, and B.J. Appelmelk, Mycobacterium marinum MMAR_2380, a predicted transmembrane acyltransferase, is essential for the presence of the mannose cap on lipoarabinomannan. *Microbiology* 156 (2010) 3492-502.

[66] D. Kaur, M.R. McNeil, K.H. Khoo, D. Chatterjee, D.C. Crick, M. Jackson, and P.J. Brennan, New insights into the biosynthesis of mycobacterial lipomannan arising from deletion of a conserved gene. *J Biol Chem* 282 (2007) 27133-40.

[67] A.K. Mishra, L.J. Alderwick, D. Rittmann, R.V. Tatituri, J. Nigou, M. Gilleron, L. Eggeling, and G.S. Besra, Identification of an alpha(1-->6) mannopyranosyltransferase (MptA), involved in Corynebacterium glutamicum lipomanann biosynthesis, and identification of its orthologue in Mycobacterium tuberculosis. *Mol Microbiol* 65 (2007) 1503-17.

[68] C.B. Sena, T. Fukuda, K. Miyanagi, S. Matsumoto, K. Kobayashi, Y. Murakami, Y. Maeda, T. Kinoshita, and Y.S. Morita, Controlled expression of branch-forming mannosyltransferase is critical for mycobacterial lipoarabinomannan biosynthesis. *J Biol Chem* 285 (2010) 13326-36.

[69] L.J. Alderwick, M. Seidel, H. Sahm, G.S. Besra, and L. Eggeling, Identification of a novel arabinofuranosyltransferase (AftA) involved in cell wall arabinan biosynthesis in Mycobacterium tuberculosis. *J Biol Chem* 281 (2006) 15653-61.

[70] N. Zhang, J.B. Torrelles, M.R. McNeil, V.E. Escuyer, K.H. Khoo, P.J. Brennan, and D. Chatterjee, The Emb proteins of mycobacteria direct arabinosylation of lipoarabinomannan and arabinogalactan via an N-terminal recognition region and a C-terminal synthetic region. *Mol Microbiol* 50 (2003) 69-76.

[71] L.J. Alderwick, G.S. Lloyd, H. Ghadbane, J.W. May, A. Bhatt, L. Eggeling, K. Futterer, and G.S. Besra, The C-terminal domain of the Arabinosyltransferase Mycobacterium tuberculosis EmbC is a lectin-like carbohydrate binding module. *PLoS Pathog* 7 (2011) e1001299.

[72] H. Skovierova, G. Larrouy-Maumus, H. Pham, M. Belanova, N. Barilone, A. Dasgupta, K. Mikusova, B. Gicquel, M. Gilleron, P.J. Brennan, G. Puzo, J. Nigou, and M. Jackson, Biosynthetic origin of the galactosamine substituent of Arabinogalactan in Mycobacterium tuberculosis. *J Biol Chem* 285 (2010) 41348-55.

[73] M. Seidel, L.J. Alderwick, H.L. Birch, H. Sahm, L. Eggeling, and G.S. Besra, Identification of a novel arabinofuranosyltransferase AftB involved in a terminal step of cell wall arabinan biosynthesis in Corynebacterianeae, such as Corynebacterium glutamicum and Mycobacterium tuberculosis. *J Biol Chem* 282 (2007) 14729-40.

[74] V. Briken, S.A. Porcelli, G.S. Besra, and L. Kremer, Mycobacterial lipoarabinomannan and related lipoglycans: from biogenesis to modulation of the immune response. *Mol Microbiol* 53 (2004) 391-403.

[75] B.J. Appelmelk, J. den Dunnen, N.N. Driessen, R. Ummels, M. Pak, J. Nigou, G. Larrouy-Maumus, S.S. Gurcha, F. Movahedzadeh, J. Geurtsen, E.J. Brown, M.M. Eysink Smeets, G.S. Besra, P.T. Willemsen, T.L. Lowary, Y. van Kooyk, J.J. Maaskant, N.G. Stoker, P. van der Ley, G. Puzo, C.M. Vandenbroucke-Grauls, C.W. Wieland, T. van der Poll, T.B. Geijtenbeek, A.M. van der Sar, and W. Bitter, The mannose cap of mycobacterial lipoarabinomannan does not dominate the Mycobacterium-host interaction. *Cell Microbiol* 10 (2008) 930-44.

[76] V. Briken, S.A. Porcelli, G.S. Besra, and L. Kremer, Mycobacterial lipoarabinomannan and related lipoglycans: from biogenesis to modulation of the immune response. Molecular *Microbiology* 53 (2004) 391-403.

[77] M. Jankute, S. Grover, A.K. Rana, and G.S. Besra, Arabinogalactan and lipoarabinomannan biosynthesis: structure, biogenesis and their potential as drug targets. *Future Microbiology* 7 (2011) 129-147.

[78] L. Nguyen, and J. Pieters, The Trojan horse: survival tactics of pathogenic mycobacteria in macrophages. *Trends Cell Biol* 15 (2005) 269-76.
[79] I. Vergne, J. Chua, S.B. Singh, and V. Deretic, Cell biology of mycobacterium tuberculosis phagosome. *Annu Rev Cell Dev Biol* 20 (2004) 367-94.
[80] I. Vergne, J. Chua, and V. Deretic, Mycobacterium tuberculosis phagosome maturation arrest: selective targeting of PI3P-dependent membrane trafficking. *Traffic* 4 (2003) 600-6.
[81] R.A. Fratti, J.M. Backer, J. Gruenberg, S. Corvera, and V. Deretic, Role of phosphatidylinositol 3-kinase and Rab5 effectors in phagosomal biogenesis and mycobacterial phagosome maturation arrest. *J Cell Biol* 154 (2001) 631-44.
[82] I. Vergne, J. Chua, and V. Deretic, Tuberculosis toxin blocking phagosome maturation inhibits a novel Ca2+/calmodulin-PI3K hVPS34 cascade. *J Exp Med* 198 (2003) 653-9.
[83] Z.A. Malik, S.S. Iyer, and D.J. Kusner, Mycobacterium tuberculosis phagosomes exhibit altered calmodulin-dependent signal transduction: contribution to inhibition of phagosome-lysosome fusion and intracellular survival in human macrophages. *J Immunol* 166 (2001) 3392-401.
[84] J.B. Torrelles, R. Knaup, A. Kolareth, T. Slepushkina, T.M. Kaufman, P. Kang, P.J. Hill, P.J. Brennan, D. Chatterjee, J.T. Belisle, J.M. Musser, and L.S. Schlesinger, Identification of Mycobacterium tuberculosis clinical isolates with altered phagocytosis by human macrophages due to a truncated lipoarabinomannan. *J Biol Chem* 283 (2008) 31417-28.
[85] L. Sweet, P.P. Singh, A.K. Azad, M.V. Rajaram, L.S. Schlesinger, and J.S. Schorey, Mannose receptor-dependent delay in phagosome maturation by Mycobacterium avium glycopeptidolipids. *Infect Immun* 78 (2010) 518-26.
[86] L.S. Schlesinger, S.R. Hull, and T.M. Kaufman, Binding of the terminal mannosyl units of lipoarabinomannan from a virulent strain of Mycobacterium tuberculosis to human macrophages. *J Immunol* 152 (1994) 4070-9.
[87] C. Astarie-Dequeker, E.N. N'Diaye, V. Le Cabec, M.G. Rittig, J. Prandi, and I. Maridonneau-Parini, The mannose receptor mediates uptake of pathogenic and nonpathogenic mycobacteria and bypasses bactericidal responses in human macrophages. *Infect Immun* 67 (1999) 469-77.

[88] L.S. Schlesinger, Macrophage phagocytosis of virulent but not attenuated strains of Mycobacterium tuberculosis is mediated by mannose receptors in addition to complement receptors. *J Immunol* 150 (1993) 2920-30.

[89] P.B. Kang, A.K. Azad, J.B. Torrelles, T.M. Kaufman, A. Beharka, E. Tibesar, L.E. DesJardin, and L.S. Schlesinger, The human macrophage mannose receptor directs Mycobacterium tuberculosis lipoarabinomannan-mediated phagosome biogenesis. *J Exp Med* 202 (2005) 987-99.

[90] J.B. Torrelles, A.K. Azad, and L.S. Schlesinger, Fine discrimination in the recognition of individual species of phosphatidyl-myo-inositol mannosides from Mycobacterium tuberculosis by C-type lectin pattern recognition receptors. *J Immunol* 177 (2006) 1805-16.

[91] I. Vergne, R.A. Fratti, P.J. Hill, J. Chua, J. Belisle, and V. Deretic, Mycobacterium tuberculosis phagosome maturation arrest: mycobacterial phosphatidylinositol analog phosphatidylinositol mannoside stimulates early endosomal fusion. *Mol Biol Cell* 15 (2004) 751-60.

[92] V. Deretic, S. Singh, S. Master, G. Kyei, A. Davis, J. Naylor, S. de Haro, J. Harris, M. Delgado, E. Roberts, and I. Vergne, Phosphoinositides in phagolysosome and autophagosome biogenesis. *Biochem Soc Symp* (2007) 141-8.

[93] H.S. Spies, and D.J. Steenkamp, Thiols of intracellular pathogens. Identification of ovothiol A in Leishmania donovani and structural analysis of a novel thiol from Mycobacterium bovis. *Eur J Biochem* 224 (1994) 203-13.

[94] N.A. Buchmeier, G.L. Newton, T. Koledin, and R.C. Fahey, Association of mycothiol with protection of Mycobacterium tuberculosis from toxic oxidants and antibiotics. *Mol Microbiol* 47 (2003) 1723-32.

[95] G.L. Newton, P. Ta, and R.C. Fahey, A mycothiol synthase mutant of Mycobacterium smegmatis produces novel thiols and has an altered thiol redox status. *J Bacteriol* 187 (2005) 7309-16.

[96] M. Hernick, Mycothiol: a target for potentiation of rifampin and other antibiotics against Mycobacterium tuberculosis. *Expert Review of Anti-infective Therapy* 11 (2013) 49-67.

[97] F. Fan, M.W. Vetting, P.A. Frantom, and J.S. Blanchard, Structures and mechanisms of the mycothiol biosynthetic enzymes. *Curr Opin Chem Biol* 13 (2009) 451-9.

[98] N. Buchmeier, and R.C. Fahey, The mshA gene encoding the glycosyltransferase of mycothiol biosynthesis is essential in Mycobacterium tuberculosis Erdman. *FEMS Microbiol Lett* 264 (2006) 74-9.

[99] D.P. Bhave, W.B. Muse, 3rd, and K.S. Carroll, Drug targets in mycobacterial sulfur metabolism. *Infect Disord Drug Targets* 7 (2007) 140-58.

[100] X. Huang, E. Kocabas, and M. Hernick, The activity and cofactor preferences of N-acetyl-1-D-myo-inosityl-2-amino-2-deoxy-alpha-D-glucopyranoside deacetylase (MshB) change depending on environmental conditions. *J Biol Chem* 286 (2011) 20275-82.

[101] F. Fan, A. Luxenburger, G.F. Painter, and J.S. Blanchard, Steady-state and pre-steady-state kinetic analysis of Mycobacterium smegmatis cysteine ligase (MshC). *Biochemistry* 46 (2007) 11421-9.

[102] M. Rawat, S. Kovacevic, H. Billman-Jacobe, and Y. Av-Gay, Inactivation of mshB, a key gene in the mycothiol biosynthesis pathway in Mycobacterium smegmatis. *Microbiology* 149 (2003) 1341-9.

[103] G.L. Newton, T. Koledin, B. Gorovitz, M. Rawat, R.C. Fahey, and Y. Av-Gay, The glycosyltransferase gene encoding the enzyme catalyzing the first step of mycothiol biosynthesis (mshA). *J Bacteriol* 185 (2003) 3476-9.

[104] M. Rawat, G.L. Newton, M. Ko, G.J. Martinez, R.C. Fahey, and Y. Av-Gay, Mycothiol-deficient Mycobacterium smegmatis mutants are hypersensitive to alkylating agents, free radicals, and antibiotics. *Antimicrob Agents Chemother* 46 (2002) 3348-55.

[105] D. Sareen, G.L. Newton, R.C. Fahey, and N.A. Buchmeier, Mycothiol is essential for growth of Mycobacterium tuberculosis Erdman. *J Bacteriol* 185 (2003) 6736-40.

[106] N.A. Buchmeier, G.L. Newton, and R.C. Fahey, A mycothiol synthase mutant of Mycobacterium tuberculosis has an altered thiol-disulfide content and limited tolerance to stress. *J Bacteriol* 188 (2006) 6245-52.

[107] X. Huang, and M. Hernick, Examination of mechanism of N-acetyl-1-D-myo-inosityl-2-amino-2-deoxy-alpha-D-glucopyranoside deacetylase (MshB) reveals unexpected role for dynamic tyrosine. *J Biol Chem* 287 (2012) 10424-34.

[108] F. Fan, and J.S. Blanchard, Toward the catalytic mechanism of a cysteine ligase (MshC) from Mycobacterium smegmatis: an enzyme involved in the biosynthetic pathway of mycothiol. *Biochemistry* 48 (2009) 7150-9.

[109] M.W. Vetting, M. Yu, P.M. Rendle, and J.S. Blanchard, The substrate-induced conformational change of Mycobacterium tuberculosis mycothiol synthase. *J Biol Chem* 281 (2006) 2795-802.

[110] M.W. Vetting, P.A. Frantom, and J.S. Blanchard, Structural and enzymatic analysis of MshA from Corynebacterium glutamicum: substrate-assisted catalysis. *J Biol Chem* 283 (2008) 15834-44.

[111] M.P. Patel, and J.S. Blanchard, Synthesis of Des-myo-Inositol Mycothiol and Demonstration of a Mycobacterial Specific Reductase Activity. *Journal of the American Chemical Society* 120 (1998) 11538-11539.

[112] C.C. Miller, M. Rawat, T. Johnson, and Y. Av-Gay, Innate protection of Mycobacterium smegmatis against the antimicrobial activity of nitric oxide is provided by mycothiol. *Antimicrob Agents Chemother* 51 (2007) 3364-6.

[113] M. Rawat, C. Johnson, V. Cadiz, and Y. Av-Gay, Comparative analysis of mutants in the mycothiol biosynthesis pathway in Mycobacterium smegmatis. *Biochem Biophys Res Commun* 363 (2007) 71-6.

[114] M. Rawat, and Y. Av-Gay, Mycothiol-dependent proteins in actinomycetes. *FEMS Microbiol Rev* 31 (2007) 278-92.

[115] J. Rengarajan, B.R. Bloom, and E.J. Rubin, Genome-wide requirements for Mycobacterium tuberculosis adaptation and survival in macrophages. *Proc Natl Acad Sci U S A* 102 (2005) 8327-32.

[116] Y. Zhang, Persistent and dormant tubercle bacilli and latent tuberculosis. *Front Biosci* 9 (2004) 1136-56.

[117] K.P. Bzymek, G.L. Newton, P. Ta, and R.C. Fahey, Mycothiol import by Mycobacterium smegmatis and function as a resource for metabolic precursors and energy production. *J Bacteriol* 189 (2007) 6796-805.

[118] I. Dalle-Donne, R. Rossi, D. Giustarini, R. Colombo, and A. Milzani, S-glutathionylation in protein redox regulation. *Free Radic Biol Med* 43 (2007) 883-98.

[119] G.L. Newton, S.S. Leung, J.I. Wakabayashi, M. Rawat, and R.C. Fahey, The DinB superfamily includes novel mycothiol, bacillithiol, and glutathione S-transferases. *Biochemistry* 50 (2011) 10751-60.

[120] M. Rawat, M. Uppal, G. Newton, M. Steffek, R.C. Fahey, and Y. Av-Gay, Targeted mutagenesis of the Mycobacterium smegmatis mca gene, encoding a mycothiol-dependent detoxification protein. *J Bacteriol* 186 (2004) 6050-8.

[121] G.L. Newton, Y. Av-Gay, and R.C. Fahey, A novel mycothiol-dependent detoxification pathway in mycobacteria involving mycothiol S-conjugate amidase. *Biochemistry* 39 (2000) 10739-46.

[122] K. Van Laer, L. Buts, N. Foloppe, D. Vertommen, K. Van Belle, K. Wahni, G. Roos, L. Nilsson, L.M. Mateos, M. Rawat, N.A.J. van Nuland, and J. Messens, Mycoredoxin-1 is one of the missing links in the oxidative stress defence mechanism of Mycobacteria. *Molecular Microbiology* (2012) n/a-n/a.

[123] S. Boonyarattanakalin, X. Liu, M. Michieletti, B. Lepenies, and P.H. Seeberger, Chemical Synthesis of All Phosphatidylinositol Mannoside (PIM) Glycans from Mycobacterium tuberculosis. *Journal of the American Chemical Society* 130 (2008) 16791-16799.

[124] T. Schneider, and H.G. Sahl, Lipid II and other bactoprenol-bound cell wall precursors as drug targets. *Curr Opin Investig Drugs* 11 (2010) 157-64.

[125] D. Barkan, Z. Liu, J.C. Sacchettini, and M.S. Glickman, Mycolic acid cyclopropanation is essential for viability, drug resistance, and cell wall integrity of Mycobacterium tuberculosis. *Chem Biol* 16 (2009) 499-509.

[126] P. Saravanan, H. Avinash, V.K. Dubey, and S. Patra, Targeting essential cell wall lipase Rv3802c for potential therapeutics against tuberculosis. *J Mol Graph Model* 38 (2012) 235-42.

[127] B.A. Wolucka, M.R. McNeil, E. de Hoffmann, T. Chojnacki, and P.J. Brennan, Recognition of the lipid intermediate for arabinogalactan/arabinomannan biosynthesis and its relation to the mode of action of ethambutol on mycobacteria. *J Biol Chem* 269 (1994) 23328-35.

[128] K. Mikusova, H. Huang, T. Yagi, M. Holsters, D. Vereecke, W. D'Haeze, M.S. Scherman, P.J. Brennan, M.R. McNeil, and D.C. Crick, Decaprenylphosphoryl arabinofuranose, the donor of the D-arabinofuranosyl residues of mycobacterial arabinan, is formed via a two-step epimerization of decaprenylphosphoryl ribose. *J Bacteriol* 187 (2005) 8020-5.

[129] A. Telenti, W.J. Philipp, S. Sreevatsan, C. Bernasconi, K.E. Stockbauer, B. Wieles, J.M. Musser, and W.R. Jacobs, Jr., The emb operon, a gene cluster of Mycobacterium tuberculosis involved in resistance to ethambutol. *Nat Med* 3 (1997) 567-70.

[130] A.E. Belanger, G.S. Besra, M.E. Ford, K. Mikusova, J.T. Belisle, P.J. Brennan, and J.M. Inamine, The embAB genes of Mycobacterium avium encode an arabinosyl transferase involved in cell wall arabinan

biosynthesis that is the target for the antimycobacterial drug ethambutol. *Proc Natl Acad Sci U S A* 93 (1996) 11919-24.

[131] M. Protopopova, C. Hanrahan, B. Nikonenko, R. Samala, P. Chen, J. Gearhart, L. Einck, and C.A. Nacy, Identification of a new antitubercular drug candidate, SQ109, from a combinatorial library of 1,2-ethylenediamines. *J Antimicrob Chemother* 56 (2005) 968-74.

[132] K. Tahlan, R. Wilson, D.B. Kastrinsky, K. Arora, V. Nair, E. Fischer, S.W. Barnes, J.R. Walker, D. Alland, C.E. Barry, 3rd, and H.I. Boshoff, SQ109 targets MmpL3, a membrane transporter of trehalose monomycolate involved in mycolic acid donation to the cell wall core of Mycobacterium tuberculosis. *Antimicrob Agents Chemother* 56 (2012) 1797-809.

[133] V. Makarov, O.B. Riabova, A. Yuschenko, N. Urlyapova, A. Daudova, P.F. Zipfel, and U. Mollmann, Synthesis and antileprosy activity of some dialkyldithiocarbamates. *J Antimicrob Chemother* 57 (2006) 1134-8.

[134] T. Christophe, M. Jackson, H.K. Jeon, D. Fenistein, M. Contreras-Dominguez, J. Kim, A. Genovesio, J.P. Carralot, F. Ewann, E.H. Kim, S.Y. Lee, S. Kang, M.J. Seo, E.J. Park, H. Skovierova, H. Pham, G. Riccardi, J.Y. Nam, L. Marsollier, M. Kempf, M.L. Joly-Guillou, T. Oh, W.K. Shin, Z. No, U. Nehrbass, R. Brosch, S.T. Cole, and P. Brodin, High content screening identifies decaprenyl-phosphoribose 2' epimerase as a target for intracellular antimycobacterial inhibitors. *PLoS Pathog* 5 (2009) e1000645.

[135] P.A. Frantom, J.K. Coward, and J.S. Blanchard, UDP-(5F)-GlcNAc acts as a slow-binding inhibitor of MshA, a retaining glycosyltransferase. *J Am Chem Soc* 132 (2010) 6626-7.

[136] B.B. Metaferia, B.J. Fetterolf, S. Shazad-Ul-Hussan, M. Moravec, J.A. Smith, S. Ray, M.T. Gutierrez-Lugo, and C.A. Bewley, Synthesis of natural product-inspired inhibitors of Mycobacterium tuberculosis mycothiol-associated enzymes: the first inhibitors of GlcNAc-Ins deacetylase. *J Med Chem* 50 (2007) 6326-36.

[137] D.W. Gammon, D.J. Steenkamp, V. Mavumengwana, M.J. Marakalala, T.T. Mudzunga, R. Hunter, and M. Munyololo, Conjugates of plumbagin and phenyl-2-amino-1-thioglucoside inhibit MshB, a deacetylase involved in the biosynthesis of mycothiol. *Bioorg Med Chem* 18 (2010) 2501-14.

[138] M.T. Gutierrez-Lugo, and C.A. Bewley, Susceptibility and mode of binding of the Mycobacterium tuberculosis cysteinyl transferase

mycothiol ligase to tRNA synthetase inhibitors. *Bioorg Med Chem Lett* 21 (2011) 2480-3.

[139] M.T. Gutierrez-Lugo, H. Baker, J. Shiloach, H. Boshoff, and C.A. Bewley, Dequalinium, a new inhibitor of Mycobacterium tuberculosis mycothiol ligase identified by high-throughput screening. *J Biomol Screen* 14 (2009) 643-52.

[140] G.L. Newton, N. Buchmeier, J.J. La Clair, and R.C. Fahey, Evaluation of NTF1836 as an inhibitor of the mycothiol biosynthetic enzyme MshC in growing and non-replicating Mycobacterium tuberculosis. *Bioorg Med Chem* 19 (2011) 3956-64.

[141] G.M. Nicholas, G.L. Newton, R.C. Fahey, and C.A. Bewley, Novel bromotyrosine alkaloids: inhibitors of mycothiol S-conjugate amidase. *Org Lett* 3 (2001) 1543-5.

[142] G.M. Nicholas, L.L. Eckman, S. Ray, R.O. Hughes, J.A. Pfefferkorn, S. Barluenga, K.C. Nicolaou, and C.A. Bewley, Bromotyrosine-derived natural and synthetic products as inhibitors of mycothiol-S-conjugate amidase. Bioorg Med Chem Lett 12 (2002) 2487-90.

[143] S.K. Kottakota, D. Evangelopoulos, A. Alnimr, S. Bhakta, T.D. McHugh, M. Gray, P.W. Groundwater, E.C. Marrs, J.D. Perry, C.D. Spilling, and J.J. Harburn, Synthesis and biological evaluation of purpurealidin E-derived marine sponge metabolites: aplysamine-2, aplyzanzine A, and suberedamines A and B. *J Nat Prod* 75 (2012) 1090-101.

[144] B. Fetterolf, and C.A. Bewley, Synthesis of a bromotyrosine-derived natural product inhibitor of mycothiol-S-conjugate amidase. *Bioorg Med Chem Lett* 14 (2004) 3785-8.

[145] B.B. Metaferia, S. Ray, J.A. Smith, and C.A. Bewley, Design and synthesis of substrate-mimic inhibitors of mycothiol-S-conjugate amidase from Mycobacterium tuberculosis. *Bioorg Med Chem Lett* 17 (2007) 444-7.

In: Inositol
Editors: H. Rocha and M. Cardoso
ISBN: 978-1-62808-520-4

Chapter 3

Inositols and Epilepsy

***R. Solomonia,*[1,2] *E. Lepsveridze,*[1]**
***T. Kiguradze*[2] *and M. Nozadze*[1,2]**
[1]Institute of Chemical Biology, Ilia State University, Tbilisi, Republic of Georgia
[2]I. Beritashvili Center of Experimental Biomedicine, Tbilisi, Republic of Georgia

Abstract

Epilepsy is a heterogeneous group of disorders. It is the most common neurological disorder after stroke. Despite achieved progress in the treatment of epilepsy, about one third of the patients with epilepsy are resistant to existing pharmacotherapies. The current treatment of epilepsy focuses exclusively on preventing or suppressing seizures, which are the symptoms of the underlying disease.

The most important challenge is to prevent epileptogenesis, the process by which brain becomes epileptic. Drugs that prevent the development of epilepsy are not yet available.

The central nervous system is characterized with relatively high concentrations of myo-inositol (MI) as well as the means to synthesize it. MI serves not only as a precursor molecule for inositol lipid synthesis, but also as a physiologically important osmolyte. Alterations in MI deposition may playrole in a number ofneuropathological conditions, either as a physiologically important osmolyte or as a precursor molecule for phosphoinositide synthesis.

Several lines of evidence indicate the involvement of changes in inositols and other osmolytes in epilepsy and the effects of inositols in the regulation of induced seizures. We have revealed that the water extract of the medicinal plant *Aquilegia vulgaris* (a plant widely used in Oriental folk medicine as antiepileptic and soporific treatments) contains compounds altering binding of ligands to the benzodiazepine andgamma-aminobuturic acid (GABA) binding sites of the GABA-A receptors. We have identified two such compounds of this extract: (1) (MI) and (2) sleep-inducing lipid oleamide. Further we have shown that MI and scyllo-inositol (SCI) pretreatment significantly decreases the severity of seizures induced either by pentylentetrazolium or kainic acid (KA). As these effects were achieved by physiological concentrations of the administered inositols, we have hypothesized that MI and SCI could represent endogenous anti-convulsants.

Our original data showed that MI also could interfere with the process of epileptogenesis. We have first induced status epilepticus by KA and then applied MI daily treatment. We have shown that a 28-day MI treatmentsignificantly attenuates biochemical changes associated with the process of epileptogenesis in the hippocampus, and restores the amount of some drastically reduced proteins to the normal level. To these proteins belongs the GLUR1 subunit of glutamate receptors, calcium/calmodulin dependent protein kinase II (CaMKII), a $\gamma 2$ subunit of GABA-A receptors. The obtained results indicate that MI treatment could at least modify the epileptogenesis process induced bybrain insult. Further studies of MI and SCI action could lead to more successful translational research and development of inositols as future anti-epileptic compounds.

1. Introduction

Epilepsy is a common and diverse set of chronicneurological disorders characterized by seizures and the cognitive, psychosocial, and social consequences of this condition (for a recent review see Loescher and Brandt, 2010).It is the most common neurological disorder after stroke, and there is a 2–3% lifetime risk of being given a diagnosis of epilepsy (Browne and Holmes, 2001). Some definitions of epilepsy require that seizures be *recurrent* and *unprovoked* (Chang and Lowenstein, 2003), but others require only a single seizure combined with brain alterations which increase the chance of future seizures (Fisher et al., 2005).Seizures can be divided into two main groups 1. Partial (Focal or local) seizures and 2.Generalized (Convulsive or non convulsive) seizures. In each of these groupsseveral subtypes of seizures are observed (Chang and Lowenstein, 2003)

The process by which a normal brain becomes epileptic is called epileptogenesis. It is a seizure-free interval lasting months to years between brain insults and the onset of symptomatic epilepsy (McNamaraet al., 2006;Jacobs et al., 2009).Historically, it was generally believed that epilepsy came about through a supernatural process. Even within the medical profession, it was not until the 18thcentury that ideas of epileptogenesis as a supernatural phenomenon were abandoned. However, biological explanations have also long existed, and sometimes explanations contained both biological and supernatural elements (Eadie and Bladin, 2001).Nowadays the widely accepted hypothesis holds that there is a cascade of poorly understood changes that transform the nonepileptic brain into one that generates spontaneous recurrent seizures (Jacobs et al., 2009; Pitkänen, 2010; Walker et al., 2002). Inepileptogenesis, an array of events occurs on molecular and cellular levels and causes neurons to fire in a disordered manner. Numerous possible mechanisms underlying this process have been suggested, but no consensus has emerged about which of the observed changes is causal or consequential and how they interact (Loescher and Brandt, 2010)

It is important to note that the brain tries to repair itself after damage, which may contribute to the fact that only a fraction of patients develop epilepsy after brain insults. It is thus vital to understand which of the molecular and cellular alterations induced by brain insults contribute directly to the development of epilepsy and which are involved in the attempt of the brain to repair the damage and recover lost function (Dichter, 2009; Jacobs et al., 2009). The latent period after brain insults may offer a window of opportunity in which an appropriate treatment may stop or modify the epileptogenic process induced by a brain insult (Jensen, 2009; Pitkanen, 2004).

Despite the progress in the pharmacotherapy of epilepsy, about one-third of epileptic patients are resistant to current treatments (Löscher et al.,2008; Sasa, 2006). Available therapy is only symptomatic and often ineffective (Löscher et al., 2008; Pitkänen, 2002). At present there is no anti-epileptic drug (AED) which could effectively prevent or modify epileptogenesis. Thus the search for a truly anti-epileptogenic drug is a problem of outstanding importance for modern biomedical sciences.

Several medicinal plant extracts have been tried and some of them are used as epilepsy treatments (e.g. for review seeKumar et al., 2012).

More than 15 years ago our attention became focused on a medicinal plant *Aqulegia vulgaris,* which belongs to genus *Aquilegia* and was widely used in

Chinese and Tibetan folk medicine as antiepileptic and soporific treatments (Shreter, 1975). We have hypothesized that the water extract of this plant contains a compound acting on gamma-aminobutiric acid type A (GABA-A) receptors and we have tested it experimentally *in vitro*. Indeed we have shown that low-molecular weight (<1Kda) aqueous fraction of this plant increases nearly two-fold the binding of 3H-flunitrezepam(agonist for benzodiapezine site of GABA-A receptors) and completely abolishes the binding of 3H-muscimol (GABA-A receptor agonist) to rat brain membranes *in vitro* (Solomonia et al., 2004).

Two rounds of High performance liquid chromatography (HPLC) and Gas chromatography/mass spectrometry detection revealed two compounds from the extract of*Aquilegia vulgaris* that influenced the binding of specific ligands with GABA-A receptors. These compounds were *myo*-inositol (MI)and oleamide-sleep inducing lipid (Solomonia et al., 2004).

Oleamide was originally isolated from the cerebrospinal fluid of sleep-deprived cats and was named as a sleep-inducing lipid (Cravatt et al., 1995). Sleep induction time is reduced by oleamide without altering the duration of rapid eye movement sleep (Basile et al., 1999; Mendelson and Basile, 1999). Oleamide is environmentally ubiquitous, being found in a variety of vegetable oils, and is used as an industrial lubricant in polyolefin manufacturing (Molnarand Erucamide1974). We have shown that oleamide exhibits antiepileptic activity and significantly decreases the degree of pentylenetetrazole-induced seizures (Solomonia et al., 2008).

The inositols are the nine isomeric forms of cyclohexanehexol, a group of small and chemically stable polar molecules that have versatile properties (Fisher et al., 2002; Michell et al., 2008).MI is the most abundant isomer and is a ubiquitous component of all eukaryotic cells. The central nervous system (CNS) is characterized with relatively high concentrations of MI as well as the means to synthesize it. MI serves not only as a precursor molecule for inositol lipid synthesis, but also as a physiologically important osmolyte (Fisher et al., 2002).MI could be directly up-taken from extracellular spaces by the specific sodium/myo-inositol cotransporter (SMIT) and H+/myo-inositol cotransporter (HMIT). In addition to this mechanism mammalian cells have two pathways by which they synthesize MI. The first is by recycling MI in a multistep dephosphorylation of inositol polyphosphate species, such as inositol trisphosphate, liberated from membrane-associated phosphatidylinositol 4,5-bisphosphate. The second is through de novo synthesis from glucose phosphate, producing MI monophosphate.

The final step in both biochemical pathways is the dephosphorylation of MI monophosphate, which is mediated by myo-inositol monophosphatase (IMPase: EC 3.1.3.25) and free MI is produced (reviewed in Fisher et al., 2002)

Alteration in MI deposition may play a role in the number of neuropathological conditions, where MI acts either as a physiologically important osmolyte or as a precursor molecule for phosphoinositide synthesis (Fisher et al., 2002).

For instance, at least part of the pathogenesis of Down's syndrome was supposed to be related to the function of inositol as an osmolyte in the nervous system. Band q22 of the human chromosome 21, which contains the gene encoding the sodium *myo*-inositol transporter (SMIT), is triplicated (Berry et al., 1995) in Down's syndrome patients and the inositol concentrations in the brain and cerebrospinal fluid (CSF) are 30–50% higher than in healthy controls (Berry, et al., 1999; Shetty, et al., 1995).

Contrary to Down's syndrome, the CSF of both unipolar and bipolar depressed patients are significantly lower in concentrations of inositol as compared to healthy control subjects (Barkai, et al., 1978). Inositol is widely used for the treatment of human neuropsychiatric disorders such as depression (Levine et al., 1995), panic disorder (Benjamin et al., 1995), and obsessive-compulsive disorder (Fux et al., 1996).

The presence ofMI in the plant extracts that possessed anti-epileptic properties prompted us to study the role of inositol in epilepsy. Our data provide ample evidence indicating that inositol could be involved in the treatment of epilepsy. Below we summarize and discuss the data about the involvement of inositol in epilepsy.

2. Changes in Metabolism of Inositols and Epilepsy

Changes in MI metabolism could be involved in different types of epilepsy, and are observed in various animal epilepsy models as well as in humans.

Temporal lobe epilepsy (TLE) patients are characterized with MI abnormalities; MI changes distinguish between the seizure focus, where MI is increased, and areas of seizure spread where MI is decreased.

The temporal lobe,ipsilateral to seizure origin in hippocampal sclerosis TLE (HS-TLE), but not late onset TLE (LO-TLE),had reduced N-acetylaspartate (NA) and elevated MI levels. Frontal lobe MI level was low in both patient groups. Ipsilateral frontal lobes had lower MI than contralateral frontal lobes (Wellard et al., 2003).

Research done on identification of novel loci and genes associated with susceptibility to febrile seizure (FS) revealed that the gene coding for MI monophosphatase (IMPase) 2 is likely to be the FS susceptibility gene. This study included participants that were the FS probands and family members of 59 Japanese nuclear families (223 members including 112 affected children). Forty-eight of these families had at least two affected children for which genome-wide linkage screening was carried out. Obtained data suggested linkage of FSs to chromosome 18p11.2. This region includes the IMPA2 gene, which encodes MI monophosphatase 2 (Nakayama et al., 2004). In the phosphatidylinositol-signaling pathway, IMPase is inhibited by lithium, which has a proconvulsant effect, and is stimulated by carbamazepine, an anticonvulsant.

The role of the IMPA2 gene was also studied in a cohort of 96 unrelated Caucasian subjects with a history of FS (Blair et al., 2007). No significant differences in genotypes of cases and matched controls were identified; no mutations or non-synonymous polymorphisms were detected in these individuals. These data suggest that the genetic variants in the IMPA2 gene are not associated with a risk of FS in Caucasian patients and patients from various genetic groups are likely to have different genetic causes of FS (Blair et al., 2007).

Kainic acid (KA) induced status epilepticus (SE)is followed by time-dependent changes in SMIT gene expression (Nonaka et al., 1999). SMIT mRNA began to increase in the brain 2 h after onset of seizure, and peaked at 12 h. In situ hybridization revealed a rapid increase of SMIT mRNA (2 h of seizure) in the CA3 hippocampal pyramidal cells and in the dentate granular cells. Then, at 4-6 h SMIT mRNA expression was observed in the other limbic structures such as the amygdala and piriform cortex. Finally, in the neocortex and in CA1 pyramidal cells, SMIT mRNA was slowly increased and peaked at 12 h. Microautoradiogram demonstrated that cells expressing SMIT mRNA were mainly neurons. These results suggest that SMIT mRNA is upregulated by KA-induced seizure primarily in structures involved in seizure activity (Nonaka et al., 1999).

MI is not the onlymolecule with osmolyte features that is involved in epilepsy.

Betaine is an organic osmolyte whichhas been suggested to be a protective osmolyte in the brain (see e.g. Bitoun and Tappaz 2000). Betaine is transported together with GABA by betaine/GABA transporter (BGT1) (Yamauchi et al., 1991; Matskevitchet al., 1999). KA induced SE is followed by strong upregulation of BGT1in glial cells at 1 week after kainate injection and with decline 3 weeks post KA injection (Zhu and Ong, 2004). Opposite to the increase in BGT-1 1 week after KA treatment, a decrease in the amounts of GABA transportersGAT-1 and GAT-3 were detected, whereas no changes were observed for GAT-2. Specific changes in expression of BGT-1 might result in alterations in the levels of GABA/betaine in the extracellular space, with consequent effects on neuronal excitability or osmolarity (Zhu andOng, 2004).

Changes in another small organic osmolyte – taurine is also associated with epilepsy. Taurine (2-aminoethanesulfonic acid) is a β-amino acid, which is not incorporated into proteins and exists in a free condition in the body. Taurine is involved in osmoregulation (Nagelhus et al., 1993, Law, 1998), membrane stabilization (Wright et al., 1986) and neuroprotection from excitotoxic cell death (Huxtable 1992).Taurine deficiency causes the prolongation of seizure activities and the persistence of a state of hyperexcitability in the brain (Mutani et al., 1977). In the human brain, the taurine level is lower than normal in the most hyperexcitable areas (Van Gelder et al., 1972). Six months after KA induced SE in the group of rats with strong behavioral response, increased taurine levels were found in the hippocampus and caudate nucleus. It is suggested that increased taurine levels in the hippocampus may involve processes for membrane stabilization, thus favoring recovery after neuronal hyperactivity (Baran 2006).

Pilocarpine induced SE is also followed by the time-dependent changes in osmoprotective genes expression (Rowley et al., 2011). Expression of BGT1, SMIT and taurine transporter werealready increased 24h after SE. From these osmoprotective genes, SMIT upregulated expression is the strongest and is considered long-term (Rowley et al., 2011).

3. Influence of Inositolson Acute Seizures

Lithium pretreatment in rats potentiates the epileptogenic effects of pilocarpine and other cholinergic agonists (Kofman et al., 1993).Lithium inhibits the enzyme inositol monophospatase and could act through a decreasedMI level (inositol depletion hypothesis Berridge et al., 1989).

This assumption was the main reason for studying MI influence on a lithium-pilocarpine model of seizures. The anti-convulsant potential of MI has been shown in experimentson lithium-pilocarpine induced seizures. In order to determine if this effect of lithium could be reversed by MI, rats were pretreated with intracerebroventricular (ICV) injections of MI, artificial CSF or L-chiro-inositol. Lithium chloride, 3 mg/kg was administered intraperitoneally 20-24 h prior to the subcutaneous injection of pilocarpine, (20 or 30 mg/kg). At both doses of pilocarpine, MI significantly prolonged the latency to the appearance of clonic seizures and lowered the pilocarpine induced seizure score (Kofman et al., 1993). MI prevented the development of clonic seizures in 50% of the rats receiving pilocarpine, 20 mg/kg. The levels of cortical MI in rats injected with MI were approximately double those of the CSF and L-chiro-inositol groups (Kofman et al., 1993). In another series of experiments besides pilocarpine, the agonist for 5HT2/5HT1c serotonergic receptor (DOI) was included and the hippocampal and cortical EEG was recorded. Administration of DOI (8 mg/kg) or pilocarpine (30 mg/kg) to rats pretreated with lithium acutely (3 mmol/kg) or chronically (dietary, 4 weeks) resulted in seizures, whereas these doses did not cause seizures without lithium pretreatment. This indicated that lithium most likely affects a signal transduction process common to both systems, which is the phosphoinositide second messenger system. To examine the potential influence of altered inositol levels on these responses, the effects of infusions (10 mg, i.c.v.) of MI, and of epi-inositol 30 min before pilocarpine or DOI were tested. Administration of MI slightly reduced the incidence of seizures induced by acute lithium plus DOI but almost completely blocked seizures induced by acute lithium plus pilocarpine. Epi-inositol also blocked seizures under both conditions but it was less effective than MI after treatment with acute lithium plus pilocarpine. Epi-inositol is an inositol isomer which is not used for phosphoinositide synthesis. Thus these data indicated that a mechanism other than phosphoinositide repletion contributed to the response of MI and epi-inositol (Williams and Jope 1995).In the same series of experiments peripheral administration of a high dose of MI blocked seizures induced by acute lithium plus pilocarpine, but the inositol treatment itself was toxic and caused seizures prior to pilocarpine administration, so the mechanism of action cannot simply be attributed to increased brain inositol levels (Williams and Jope 1995).Existence of MI in the water extract of *Aquilegia vulgaris,* its effects on 3H-muscimol binding in vitro (Solomonia et al., 2004) and the above discussed data prompted us to carry out a special series of experiments of MI effects on acute seizures.

The rational basis of our design was to use intraperitoneal (i.p.)injections of MI in the range of physiological concentrations and test the effects of the inositols on chemically induced seizures according to the classically approved schemes, e.g. 30 minute before convulsant application (see Vogel 2002).In the first series of experiments we used two types of chemical convulsants: pentilentetrazolium (PTZ) and KA. MI in the dose of 20 mg/kg was injected i.p. 30 minutes before chemical convulsant administration. MI pretreatment induced a significant decrease in the seizure score for both types of chemical convulsant induced seizures (Solomonia et al., 2007).

In the next series of experiments we focused only on PTZ induced convulsions, but more parameters were scored and more compounds were tested (Nozadze et al., 2011).In the control group design we added a group treated with mannitol – polyol of the same molecular weight as inositols. Besides MI, the effect of another inositol –scyllo-inositol (SCI) was investigated. SCI is present in the human brain in concentrations estimated to be between 5 to 12% of MI[MI concentration in the rat brain is in the range of 7-11.5 mmol/kg wet wt, see review Fisher et al., 2002].In rat and rabbit brains, MI is converted into SCI by specific epimerase (Sherman et al., 1968, Fisher et al., 2002).The existence of such specific enzyme for SCI synthesis suggests SCI's important physiological role in the brain.

Half an hour after treatment with MI (30 mg/kg) or SCI (5 mg/kg) seizures were induced in Wistar rats with PTZ (60 mg/kg). Control animals were treated either by normal saline or mannitol (30 mg/kg), given at the same time interval before PTZ injection, as MI/SCI groups. The anticonvulsant effects of MI/SCI treatment were assessed by the latent period (the time from PTZ-injection to the onset of first seizures), and the duration and severity (score) of seizures. Both MI and SCI treatment significantly reduced the seizure score and seizure duration as compared to control groups treated either by saline or mannitol (Nozadze et al., 2011).Both inositiols effectively increased the latency of seizure onset. Inositol pretreatment did not change the mortality rate (Nozadze et al., 2011).

Despite a six times higher concentration of MI compared to SCI in these experiments, no significant differences were observed between MI+PTZ and SCI+PTZ groups. Inositol stereoisomers have a different distribution of hydroxyl groups across the surfaces of the sugar ring and these differences could account for different effective concentrations of MI and SCI on PTZ induced seizures.

The effects of MI and SCI in these experiments revealed that they are specific to MI and SCI as inositols.

No statistically significant differences were found between two sham groups: saline and mannitol treated, and both groups were significantly different from the MI and SCI groups with respect to nearlyall of the studied parameters (Nozadze et al., 2011). These results have convincingly demonstrated that MI and SCI, in the range of physiological concentrations, possess anticonvulsant properties on PTZ induced convulsions. Based on these and some of our other results we have proposed that, in addition to other functions, MI and SCI are endogenous anticonvulsants (Nozadze et al., 2011).

Other osmolytes of the organism also revealed anti-seizure activities. Betaine was found to block the induction of convulsions by electroconvulsive shock and by PTZ (Freed et al., 1979). Taurine displays anti-convulsive activitiesin the model of ouabain-induced seizures in rats (Barbeau and Donaldson 1974). Taurine pretreatment (150 mg/kg) before KA application significantly reduces convulsions for the following 2 hours and has neuroprotective effects. KA induced molecular effects, measured on the third day after SE, also disappear with taurine pretreatment (Junyent et al., 2009).

4. Inositols and Epileptogenesis

Ourexperiments and those by other authors'(above) have described experiments where inositol or other osmolyte pretreatments were applied before chemical convulsant induced brain insult (e.g. Solomonia et al., 2007, Nozadze et al., 2011, Junyent et al., 2009, Freed et al., 1979, Williams and Jope 1995). Such design is not applicable for testing inositols as an antiepileptogenic compound (see Loesher and Brandt 2010). As it was demonstrated inositol pretreatment attenuates the severity and shortens the duration of SE (Solomonia et al., 2007, Nozadze et al., 2011)and thereby could reduce the long-term consequences of the brain insult.

Therefore in the next series of experiments we have first induced the SE by KA, and then tried MI daily treatment. With such design of experiment antiepileptogenic properties of the compound could be explored. It is clear that treatment strategies that could interfere with the process of epileptogenesis would have significant benefits over the current approach of AEDs (see e.g, Loscher and Brandt, 2010) and will be of great importance for epilepsy treatment.

The diagram of our behavioral and biochemical experimental design is provided on Figure1. Briefly, experiments were carried out on four groups of Wistar rats.

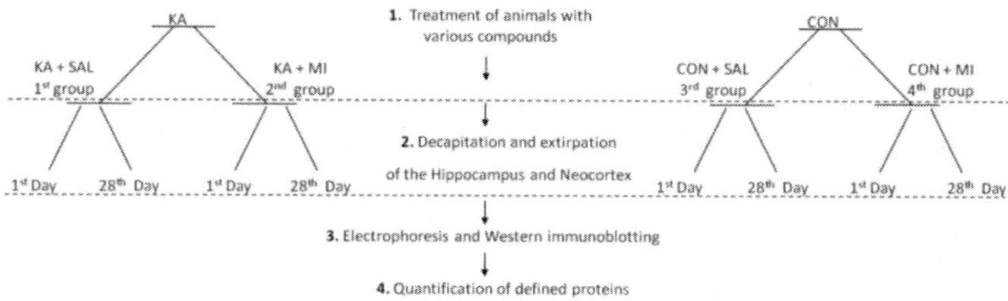

Figure 1.General diagram of experimental design.

Each group was divided in two subgroups according to the time of decapitation (the first or the twenty-eighth day after the start of the experiment, see Figure1).Six hours following KA treatment, the first group received a saline injection (KA+SAL group), while the second group received an MI (30 mg/kg) injection (KA+MI group). The next morning, half of the KA+SAL group received a saline injection, while half of the KA+MI group received MI injections.

Animals from these subgroups were killed 10 h afterwards [time corresponding to the 28–30 h following KA treatment (first day – Figure1). Remaining rats were treated twice daily either by saline (KA+SAL group) or MI (KA+MI group) during the following 27 days and then decapitated (28th day of experiment).

Spontaneous seizures in KA-treated animals during the first month are rare, but could take place prior to the twenty-eighth day (Williams et al.,2009). If such seizures take place a few days before the second time-point decapitation, it is possible that the detected changes are the consequence of acute seizure, but not of the long-term process of epileptogenesis.

Therefore, starting from the twenty-fourth day of the experiment, the rats were monitored by an infrared video system for 24 h a day. The third and fourth groups of animals were control groups and were not treated by KA. Rats in the third group received saline (CON+SAL), while the fourth group received MI injections (30 mg/kg, CON +MI) twice a day. They were killed according to the same time schedule as groups treated with KA.

After decapitation, two regions - hippocampus and neocortex- were removed from each brain. The electrophoresis and quantitative Western Immunoblotting of nuclear free homogenate fractions from obtained tissue samples were carried as described in our work (Solomonia et al., 2010, 2013).

In the first series of experiments we focused on the changes of the following proteins: α-subunit of calcium-calmodulin-dependent protein kinase II (αCaMKII); (ii) GLUR1 subunit of α-amino-3-hydroxy-5-methyl-4-isoxazole propionate (AMPA) type glutamate receptors (GLUR1) and (iii) heat shock protein 90 (HSP90). The reasons for choosing these proteins were the following: (i) there is a strong evidence that decreased expression of αCaMKII is sufficient to induce limbic epilepsy(Butler et al., 1995; Churn et al., 2000; reviewed in McNamara et al., 2006); (ii) the amount of GLUR1 immunopositive neurons is known to be drastically reduced after SE (Tang et al., 2005); (iii) whereas the amount of HSP90 is not changed (Gass et al., 1995).

The amount of neither protein was changed significantly by experimental conditions in the hippocampus or neocortex1 day after beginning theexperiment or at the 28th day in the neocortex.

The mean amount of the GLUR1 subunit was drastically reduced (more than 95%) in the hippocampus of the KA+SALgroup of rats and was significantly lower as compared to two groups of controlrats. The mean amount ofGLUR1 AMPA receptors in the KA+MI group was in the middle between the KA and control groups: significantly lower as compared to two control groups but significantly higher as compared to the KA group (Figure2). Thus MI treatment significantly halts the drastic reduction of GLUR1 AMPA receptors induced by KA evoked SE (Solomonia et al., 2010).

The changes for CaMKII in general were similar to those observed for the GLUR1 AMPA receptor. The mean amount of CaMKII in the KA+Sal group was significantly decreased (by approximately 60%) as compared to two control groups.

The mean amount of CaMKII in the KA+MI group was significantly more as compared to the KA+SAl group,which was closer and not different from two control groups. The mean amounts in the two control groups were not different from each other (Figure2). The mean amounts of HSP90 protein did not differ significantly between the experimental groups (Figure2).Thus the results of these series of experiments revealed that long-term MI treatment can halt the biochemical processes induced by KA induced SE (Solomonia et al., 2010).

The process of epileptogenesis is associated with changes in the expression and function of GABA-A receptors (GABA-AR)(Friedman et al. 1994; Kharlamov et al. 2011; Sperk et al. 1998; for a review see GonzalezandBrooks-Kayal 2011).

Various drugs that enhance GABAergic inhibition are commonly used as AEDs. In the next series of experiments we have examined the influence ofdaily MI treatment by the same design (see Figure1) on changes in GABA-AR subunit amounts.

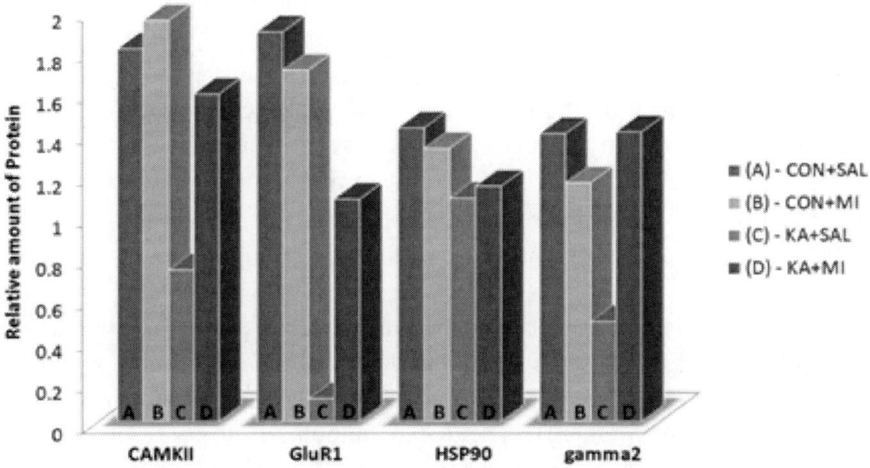

Figure 2.Mean relative amounts of CamKII, GluR1, HSP90 and gamma2 proteins in of rat brain hippocampus from four groups of rats on 28th day of experiment. Columns "A" correspond to CON+SAL group, columns "B" – to CON+MI, columns "C" to KA+SAL and columns "D" to KA+MI groups. For the details of calculations of relative amounts of proteins see Solomonia et al., 2011; 2013.

Among the numerous subunits of GABA-ARs we decided to focus on the following ones: α1, α4, γ2 and δ. The reasons for this selection were: (i) the expression of these subunits undergoes time-dependent changes during epileptogenesis and following epilepsy (Friedman et al. 1994; Gonzalez and Brooks-Kayal 2011; Sperk et al. 1998); and (ii) α1 and γ2 are found in synapses and participate in phasic inhibition (Mangan et al. 2005; Nusser and Mody 2002) whereas α4 and δ subunits are crucial for tonic inhibition and are predominantly expressed in extrasynaptic location (Jia et al. 2005; Wei et al. 2003). Both types of inhibitions are modified in epilepsy (Pavlov et al. 2011; Zhang et al., 2007).

One day after the beginning of the experiment the significant changes were observed only for the α1 subunit – the protein amount was reduced in the hippocampus of the KA+SAL and KA+MI groups as compared to the two control groups and MI did not exert any influence on KA treated animals. 28 days later the mean amount of the α1 subunit in the hippocampus was not significantly higher in the KA+SAL and KA+MI groups than in the control groups. The late increase of the α1 subunit could account for a compensatory increase of the α1 subunit containing GABA-ARs. There were no differences between KA treated groups or the two control groups.

MI treatment does not have any additional influence on this late increase of α1 subunits (Solomonia et al., 2013).

An interesting effect was observed for the α4 subunit on the 28th day after starting the experiment in the hippocampus. KA-induced SE was followed by a decrease of the mean amount of protein as compared to all three other groups. MI treatment after KA induced SE not only restored the amount of α4, but increased it significantly even compared to the control groups. No differences were observed between the control groups (Solomonia et al., 2013).

The strongest effect of MI treatment was observed for the γ2 subunitin the hippocampus. In the KA+SAL group the drastic reduction of this subunit was observed compared to the two control groups as well as compared to the KA+MI group (see Figure2). The mean amount of the γ2 subunitin the KA+MI group is not different from the CON+SAL group and is even significantly higher as compared to the CON+MI group. Thus MI treatment as in case of CaMKII and GluR1 restored the amount of the γ2subunit to the normal level. The MI effect in the hippocampus for the γ2 subunit at the 28th day was significantly different as compared to the neocortex (see Solomonia et al., 2013).γ2 subunits are important for the assembly of GABA-A receptors, their accumulation at synapses, surface stability, and synaptic clustering (Alldred et al. 2005; Essrich et al. 1998). Hence restoration of its level should normalize the function of the GABA-A receptor system. No significant changes were found for the *δ* subunit.

For the α1 subunit, significant changes were also observed in the neocortex on the 28th day of experiment. KA treatment decreased the amount of protein in both the KA+SAL and KA+MI groups as compared to the control. There was no difference between KA-treated groups or between the control groups. Thus, in the neocortex, MI has no effect on KA induced changes in the amount of the α1 subunit (Solomonia et al., 2013). The data obtained indicate that either MI effects are region-specific or MI treatment does not interfere with α1 subunit expression, neither in the necortex nor in the hippocampus.

MI pretreatment before KA induced SE exerts a strong neuroprotective effect on hippocampal cell loss during the process of epileptogenesis(Kotaria et al., 2013). As discussed above MI pretreatment evokes the decrease of KA-induced seizure severity (Solomonia et al., 2007).Thus MI treated groups were expected to have lower seizure severity grades as compared to SAL+KA treated groups. Therefore it was possible that differences revealed between the groups by hippocampal cell loss analysis would partially reflect the condition for less severe seizures in these animals. To avoid this pitfall, the rats were selected so that SAL+KA and MI+KA groups had the same seizure score.

The strongest neuroprotective effect was observed in the pyramidal layer of CA1, radial and oriental layers of CA3 fields. Ultrastructural alterations were described in CA1, 14days after treatment. The structure of neurons, synapses, and porosomes are well preserved in the rats pretreated with MI compared to rats treated with saline and then by KA (Kotaria et al., 2013)

5. Possible Mechanisms of MI Action on Acute Seizures and Epileptogenesis

What could be the mechanisms of MI action on acute seizures and on epileptogenesisprocesses? MI is the precursor of phosphoinositode biosynthesis and an initial assumption for the study of inositols influence on a lithium-pilocarpine induced seizure was “inositol-depletion hypothesis”.

However this mechanism of inositol action on acute seizures was declined, as epi-inositol, an isomer that is not used for phosphoinosotide biosynthesis, also possesses anti-convulsive features on lithium-pilocarpine induced seizures (Williams and Jope 1995).

Another mode of inositol action on seizures and epileptogenesis may be linked with its osmolytic properties.During intense neuronal excitation a massive influx of Na^+, Ca^{2+} and Cl^- takes place, which leads to inward water flow and cellular swelling (Choi 1987; Rothman 1985).

Under such conditions various enzymes cannot function in the normal way (Yanceyet al., 1993). In addition, reduced extracellular space enhances neuronal excitability (Dudek et al. 1990; Schwartzkoin et al. 1998).

To compensate for this condition neuronal and other cells increase the intracellular concentration of small organic osmolytes by upregulation of osmoprotective gene expression (e.g. Rowley et al. 2011), which restores the conditions for normal functioning of the intracellular enzymes. One such important osmolyte appears to be MI (Fisheret al. 2002). Upregulation of SMIT after KA induced seizures also indicates the beneficial effects of MI accumulation for the normalization of cellular functions (Nonaka et al., 1999). We suggest thatone of the possible mechanisms of MI action could be osmoprotection, which prevents/modifies the biochemical processes of epileptogenesis.

Based on our previous work demonstrating the influence of MI on 3H-muscimol and 3H-MK-801 binding in vitro (Solomonia et al. 2004), we additionally propose that inositols could also exert their anti-epileptic action by the modulation of neuronal activity through direct influence on GABA-ergic or/and glutamatergic synaptic transmission, in particular by modulatory action on the respective postsynaptic receptors.

Conclusion

Current evidence convincingly demonstrates that alterations in the metabolism ofinositols and some other organic osmolytes are involved in epilepsy. These molecules are able to reduce acute seizures induced by various types of convulsants. Moreover, and most importantly, MI is able to prevent/modify the process of epileptogenesis on a molecular level.

Further investigations are needed to gather more evidence about mechanisms of inositol actions which could lead to more successful translational research and development of inositols as future anti-epileptogenic compounds.

References

Alldred, M.J., Mulder-Rosi, J., Lingenfelter, S. E., Chen, G., Luscher, B. (2005) Distinct gamma2 subunit domains mediate clustering and synaptic function of postsynaptic GABA-A receptors and gephyrin. *J. Neurosci.*, 25, 594-603

Baran, H. (2006) Alterations of taurine in the brain of chronic kainic acid epilepsy model. *Amino Acids*, 31, 303–307

Barbeau, A., Donaldson, J. (1974) Zinc, taurine, and epilepsy.*Arch. Neurol.* 30, 52-8.

Barkai, A. I., Dunner, D. L., Gross, H. A., Mayo, P., and Fieve, R. R. (1978). Reduced myo-inositol levels in cerebrospinal fluid from patients with affective disorder. *Biological Psychiatry*, 13, 65–72.

Basile, A. S., Hanus, L. and Mendelson, W. B. (1999). Characterization of the hypnotic properties of oleamide. *Neuroreport*, 10, 947–51.

Benjamin, J., Levine, J., Fux, M., Aviv, A., Levy, D., and Belmaker, R. H. (1995). Double-blind, placebo-controlled, crossover trial of inositol treatment for panic disorder. *The American Journal of Psychiatry,* 152, 1084–6.

Berridge, M. J., Downes, C. P. and Hanley, M. R. (1989). Neural and developmental actions of lithium: a unifying hypothesis. *Cell*, 59, 411–9.

Berry, G. T., Mallee, J. J., Kwon, H. M., Rim, J. S., Mulla, W. R., Muenke, M., and Spinner, N. B. (1995). The human osmoregulatory Na+/myo-inositol cotransporter gene (SLC5A3): molecular cloning and localization to chromosome 21. *Genomics,* 25, 507–13.

Berry, G. T., Wang, Z. J., Dreha, S. F., Finucane, B. M., and Zimmerman, R. A. (1999). In vivo brain myo-inositol levels in children with Down syndrome. *The Journal of Pediatrics*, 135, 94–97.

Bitoun, M., Tappaz, M. (2000) Gene expression of taurine transporter and taurine biosynthetic enzymes in brain of rats with acute or chronic hyperosmotic plasma.A comparative study with gene expression of myoinositol transporter, betaine transporter and sorbitol biosynthetic enzyme.*Brain Res. Mol. Brain Res.* 77, 10 –18.

Blair, M.A., Ma, S., Abou-Khalil, B., Hedera, P. (2007) Genetic variants in the IMPA2 gene do not confer increased risk of febrile seizures in Caucasian patients. *Eur. J. Neurol.*14, 424-427.

Browne, T. R. and Holmes, G. L. (2001). Epilepsy. *The New England Journal of Medicine,* 344(15), 1145–51.

Butler, L. S., Silva, A. J., Abeliovich, A., Watanabe, Y., Tonegawa, S., and McNamara, J. O. (1995).Limbic epilepsy in transgenic mice carrying a Ca2+/calmodulin-dependent kinase II alpha-subunit mutation. *Proc. Natl. Acad. Sci. US* 92, 6852-6855

Chang, B. S. and Lowenstein, D. H. (2003). Epilepsy. *The New England Journal of Medicine*, 349(13), 1257–66.

Choi, D.W. (1987) Ionic dependence of glutamate neurotoxicity, *J. Neurosci.* 7, 369–379.

Churn, S. B., Sombati, S., Jakoi, E. R., Severt, L., DeLorenzo, R. J., and Sievert, L. (2000). Inhibition of calcium/calmodulin kinase II alpha subunit expression results in epileptiform activity in cultured hippocampal neurons. *Proc. Natl. Acad.Sci. US*97 5604–9.

Cravatt, B. F., Prospero-Garcia, O., Siuzdak, G., Gilula, N. B., Henriksen, S. J., Boger, D. L., and Lerner, R. A. (1995). Chemical characterization of a family of brain lipids that induce sleep.*Science* (New York, N.Y.), 268 (1506–9.)

Dichter, M. A. (2009). Emerging concepts in the pathogenesis of epilepsy and epileptogenesis. *Archives of Neurology,* 66, 443–7.

Dudek, F.E., Obenaus, A., Tasker, J.G. (1990) Osmolality induced changes in extracellular volume after epileptiform burst independent of chemical synapses in the rat: importance of non-synaptic mechanisms in hippocampal epileptogenesis. *Neurosci.Lett.* 120, 267–270

Eadie, M. and Bladin, P. (2001). *A Disease Once Sacred: A History of the Medical Understanding of Epilepsy*. London: John Libbey. 80

Essrich, C., Lorez, M., Benson, J.A., Fritschy, J.M., Luscher, B. (1998) Postsynaptic clustering of major GABAA receptor subtypes requires the gamma 2 subunit and gephyrin. *Nat. Neurosci.* 1, 563-571

Fisher, K.S., Novak, J.E., Agranoff, B.W. (2002) Inositol and higher inositol phosphates in neuronal tissues: homeostasis, metabolism and functional significance. *J. Neurochem.* 82, 736-54

Fisher, R. S., Van Emde Boas, W., Blume, W., Elger, C., Genton, P., Lee, P., and Engel, J. (2005). Epileptic seizures and epilepsy: definitions proposed by the International League Against Epilepsy (ILAE) and the International Bureau for Epilepsy (IBE). *Epilepsia,* 46, 470–2.

Fisher, S. K., Novak, J. E. and Agranoff, B. W. (2002). Inositol and higher inositol phosphates in neural tissues: homeostasis, metabolism and functional significance. *Journ.Neurochem.,* 82, 736–54.

Freed, W. J., Gillin, J. C. and Wyatt, R. J. (1979) Anticonvulsant properties of betaine, *Epilepsia*20, 209-213

French, E. D., Vezzani, A., Whetsell, W. O. Jr. and Schwarcz, R. (1986) Anti-excitotoxic actions of taurine in the rat hippocampus studied in vivo and in vitro. *Adv. Exp. Med. Biol.* 203, 349–362.

Friedman, L.K., Pellegrini-Giampietro, D.E., Sperber, E.F., Bennett, M.V., Moshé, S.L., Zukin, R.S.(1994) Kainate-induced status epilepticus alters glutamate and GABAA receptor gene expression in adult rat hippocampus: an in situ hybridization study. *J. Neurosci.* 14, 2697-2707.

Fux, M., Levine, J., Aviv, A., and Belmaker, R. H. (1996).Inositol treatment of obsessive-compulsive disorder. *The American Journal of Psychiatry,* 153, 1219–1221.

Gass, P., Prior, P.and Kiessling, M. (1995). Correlation between seizure intensity and stress protein expression after limbic epilepsy in the rat brain.*Neuroscience*, 65, 27–36.

González, M.I., Brooks-Kayal, A.(2011) Altered GABA(A) receptor expression during epileptogenesis. *Neurosci.Lett.* 497, 218-22

Huxtable, R.J. (1992) Physiological actions of taurine. *Physiol. Rev.*72,101–163

Jacobs, M. P., Leblanc, G. G., Brooks-Kayal, A., Jensen, F. E., Lowenstein, D. H., Noebels, J. L., Spencer, D. D., et al. (2009). Curing epilepsy: progress and future directions. *Epilepsy and Behavior* 14(3), 438–45.

Jensen, F. E. (2009). Introduction. Posttraumatic epilepsy: treatable epileptogenesis. *Epilepsia*, 50 Suppl. 2, 1–3.

Jia, F., Pignataro, S., Schofield, C.M., Yue, M., Harrison, N.L., Goldstein, P.A. (2005) An extrasynaptic GABAA receptor mediates tonic inhibition in thalamic VB neurons. *J.Neurophysiol.*94, 4491–4501

Junyent, F., Utrera, J., Romero, R., Pallas, M., Camins, A., Duque, D., and Auladell, C. (2009) Prevention of epilepsy by taurine treatments in mice experimental model.*J. Neurosci. Res.* 87, 1500–1508

Kharlamov, E.A., Lepsveridze, E., Meparishvili, M., Solomonia, R.O., Lu, B., Miller, E.R., Kelly, K.M., Mtchedlishvili, Z. (2011) Alterations of GABA(A) and glutamate receptor subunits and heat shock protein in rat hippocampus following traumatic brain injury and in posttraumatic epilepsy. *Epilepsy Res.* 95, 20-34

Kofman, O., Sherman, W. R., Katz, V., and Belmaker, R. H. (1993). Restoration of brain myo-inositol levels in rats increases latency to lithium-pilocarpine seizures. *Psychopharmacology*, 110, 229–34.

Kotaria, N., Kiladze, M., Zhvania, M.G., Japaridze, N.J., Bikashvili, T., Solomonia, R.O., Bolkvadze, T. (2013)The Protective effect of myo-inositol on hippocamal cell loss and structural alterations in neurons and synapses triggered by kainic acid-induced status epilepticus. *Cell. Mol. Neurobiology*, accepted for publication

Kumar, S., Madaan, R., Bansal, G., Jamwal, A., Sharma, A. (2012) Plants and plant products with potential anticonvulsant activity – a review. *Pharmacognosy Communications* 2, 3-99

Law, R.O. (1998) The role of taurine in the regulation of brain cell volume in chronically hyponatraemic rats. *Neurochem. Int.* 33,467–472.

Levine, J., Barak, Y., Gonzalves, M., Szor, H., Elizur, A., Kofman, O., and Belmaker, R. H. (1995).Double-blind, controlled trial of inositol treatment of depression. *The American Journal of Psychiatry*, 152, 792–794.

Loescher, W. and Brandt, C. (2010). Prevention or Modification of Epileptogenesis after Brain Insults: Experimental Approaches and Translational Research. *Pharmacol. Rev.*62, 668–700.

Löscher, W., Gernert, M.and Heinemann, U. (2008). Cell and gene therapies in epilepsy--promising avenues or blind alleys? *Trends in Neurosci.,*31, 62–73.

Mangan, P.S., Sun, C., Carpenter, M., Goodkin, H.P., Sieghart, H.P., Kapur, J. (2005) Cultured hippocampal pyramidal neurons express two kinds of GABAA receptors. *Mol.Pharmacol.*67, 775–788

Matskevitch, I.,Wagner, C.A., Stegen, C., Bro, S., Noll, B., Risler T., Kwon H.M., Handler, J.S., Waldegger, S., Busch, A.E., and Lang, F. (1999) Functional characterization of the betaine/g-aminobutyric acid transporter BGT-1 expressed in Xenopus oocytes*J.Biol.Chem.*, 274, 16709-16716

McNamara, J. O., Huang, Y. Z. and Leonard, A. S. (2006).Molecular signaling mechanisms underlying epileptogenesis. *Science's STKE*356, re:12.

Mendelson, W. B. and Basile, A. S. (1999). The hypnotic actions of oleamide are blocked by a cannabinoid receptor antagonist. *Neuroreport*, 10, 3237–3239.

Michell, R. H. (2008). Inositol derivatives: evolution and functions. *Nature Reviews. Molecular Cell Biology,* 9, 151–61.

Molnar, N.M. Erucamide. *J. Am. Oil Chem. Soc.* 51 (1974) 84–87

Mutani, R., Durelli, L., Mazzarino, M., Valentini, C., Monaco, F., Fumero, S., Mondino, A. (1977) Longitudinal changes of brain amino acid content occurring before, during and after epileptic activity. *Brain Res.* 122, 513–521

Nagelhus, E.A., Lehmann, A., Ottersen, O.P. (1993) Neuronal–glial exchange of taurine during hypo-osmotic stress: a combined immunocytochemical and biochemical analysis in rat cerebellar cortex. *Neuroscience* 54, 615–631.

Nakayama, J., Yamamoto, N., Hamano, K., Iwasaki, N., Ohta, M., Nakahara, S., Matsui, A., et al. (2004).Linkage and association of febrile seizures to the IMPA2 gene on human chromosome 18. *Neurology*, 63, 1803–1807.

Nonaka, M., Kohmura, E., Yamashita, T., Yamauchi, A., Fujinaka, T., Yoshimine, T., Tohyama, M., et al. (1999).Kainic acid-induced seizure upregulates Na(+)/myo-inositol cotransporter mRNA in rat brain. *Brain research. Molecular Brain Research,* 70, 179–86.

Nozadze, M., Mikautadze, E., Lepsveridze, E., Mikeladze, E., Kuchiashvili, N., Kiguradze, T., Kikvidze, M., Solomonia, R. (2011) Anticonvulsant activities of myo-inositol and scyllo-inositol on pentylenetetrazol induced seizures. *Seizure*,20,173-176

Nusser, Z., Mody, I. (2002) Selective modulation of tonic and phasic inhibitions in dentate gyrus granule cells. *J.Neurophysiol.* 87, 2624–2628

Pavlov, I., Huusko, N., Drexel, M., Kirchmair, E., Sperk, G., Pitkanen, A., Walker, M.C. (2011) Progressive loss of phasic, but not tonic GABA-A receptor-mediated inhibition in dentate granule cells in a model of post-traumatic epilepsy in rats. *Neuroscience* 194, 208–219

Pitkänen, A. (2002). Efficacy of current antiepileptics to prevent neurodegeneration in epilepsy models. *EpilepsyResearch*, 50(1-2), 141–60.

Pitkanen, A. (2004). New pharmacotherapy for epilepsy.*IDrugs: The Investigational Drugs Journal,* 7, 471–7.

Pitkänen, A. (2010). Therapeutic approaches to epileptogenesis--hope on the horizon. *Epilepsia*, 51 Suppl. 3, 2–17.

Rothman, S.M. (1985) The neurotoxicity of excitatory amino acid is produced by passive chloride influx, *J. Neurosci.* 5, 1483–1489.

Rowley, N.M., Smith, M.D., Lamb, J.G., Schousboe, A., and White, H.S. (2011) Hippocampal betaine/GABA transporter mRNA expression is not regulated by inflammation or dehydration post-status epilepticus. *J.Neurochem.,* 117, 82-90.

Sasa, M. (2006). A new frontier in epilepsy: novel antiepileptogenic drugs. *Journal of Pharmacological Sciences,* 100, 487–94.

Schwartzkoin, P.A., Baraban, S.C., Hochmann, D.W. (1998) Osmolarity, ionic flux and changes in brain excitability. *Epilepsy Res.*32, 275–285

Sherman, W.R., Stewart, M.A., Simpson, P.C., Goodwin, S.L. (1968) The identification of myoinosose-2 and scyllo-inositol in mammalian tissues. *Biochemistry*, 7, 819–824.

Shetty, H. U., Schapiro, M. B., Holloway, H. W., and Rapoport, S. I. (1995). Polyol profiles in Down syndrome. myo-Inositol, specifically, is elevated in the cerebrospinal fluid. *The Journal of Clinical Investigation*, 95, 542–6.

Shreter, A. I. (1975). *Medicinal Herbs of the Soviet Far East*. Moscow: Medicina, 105–106.

Solomonia, R, Kuchiashvili, N., Berulava, A., Pkhakadze, V., Trapaidze, N., Zhvania, M., and Abesadze, I. (2004). Purification and identification of components of the Aquilegia vulgaris extract fraction exhibiting anti-epileptic activity. *Journal of Biological Physics and Chemistry* 4, 185–192.

Solomonia, R., Nozadze, M., Kuchiashvili, N., Bolkvadze, T., Kiladze, M., Zhvania, M., Kigyradze, T., Pkhakadze, V.(2007) Effect of myo-inositol on convulsions induced by pentylenetetrazole and kainic acid in rats. *Bull. Exp. Biol. Med.* 143, 58-60

Solomonia, R., Nozadze, M., Mikautadze, E., Kuchiashvili, N., Kiguradze, T., Abkhazava, D., Pkhakadze, V., Mamulaishvili, I., Mikeladze, E., Avaliani, N. (2008) Effect of oleamide on pentylenetetrazole-induced seizures in rats.*Bull. Exp. Biol. Med.* 145, 225-227

Solomonia, R., Mikautadze, E., Nozadze, M., Kuchiashvili, N., Lepsveridze, E., and Kiguradze, T. (2010). Myo-inositol treatment prevents biochemical changes triggered by kainate-induced status epilepticus. *Neuroscience Letters*, 468, 277–281.

Solomonia, R., Goichaishvili, N., Nozadze, M., Lepsveridze, E., Dzneladze, D., Kiguradze.T.(2013) Myo-inositol treatment and GABA-A receptor subunit changes after kainate-induced status epilepticus. *Cell. Mol. Neurobiol.* 33, 119-127

Sperk, G., Schwarzer, C., Tsunashima, K., Kandlhofer, S. (1998) Expression of GABA(A) receptor subunits in the hippocampus of the rat after kainic acid-induced seizures. *Epilepsy Res.* 32,129-139.

Tang, F. R., Chia, S. C., Zhang, S., Chen, P. M., Gao, H., Liu, C. P., Khanna, S., et al. (2005). Glutamate receptor 1-immunopositive neurons in the gliotic CA1 area of the mouse hippocampus after pilocarpine-induced status epilepticus.*European Journal of Neuroscience*, 21, 2361–2374.

Van Gelder, N.M., Sherwin, A.L., Rasmussen, T. (1972) Amino acid content of epileptogenic human brain: Focal versus surrounding regions. *Brain Research*, 40, 385–393

Vogel, H.G. (2002) *Drug Discovery and evaluation (Pharmacological assays)* second edition, Springer, Verlag Berlin Heidelberg, pp. 422-423

Walker, M. C., White, H. S. and Sander, J. W. A. S. (2002). Disease modification in partial epilepsy. *Brain* 125(Pt 9), 1937–50.

Wei, W., Zhang, N., Peng, Z., Houser, C.R., Mody, I. (2003) Perisynaptic localization of delta subunit-containing GABA-A receptors and their activation by GABA spillover in the mouse dentate gyrus. *J. Neurosci.* 23, 10650–10661

Wellard, R. M., Briellmann, R. S., Prichard, J. W., Syngeniotis, A., and Jackson, G. D. (2003). Myoinositol abnormalities in temporal lobe epilepsy.*Epilepsia*, 44, 815–821.

Williams, M.B., Jope, R.S. (1995) Modulation by inositol of cholinergic- and serotonergic-induced seizures in lithium-treated rats. *Brain Res.* 685,169-178.

Williams, P.A., White, A.M., Clark, S., Ferraro, S.J., Swiercz, W., Staley, K.J., Dudek, F.E. (2009) Development of spontaneous recurrent seizures after kainate-induced status epilepticus. *J. Neurosci.* 29, 2103–2112

Wright, C.E., Tallan, H.H., Lin, Y.Y., Gaull, G.E. (1986) Taurine: biological update. *Annu. Rev. Biochem.* 55,427–453.

Yamauchi, A., Uchida, S., Kwon, H.M., Preston, A.S., Robey R.B., Garcia-Perez, A., Burg, M.B., Handler, J.S. (1991) Cloning of a Na+- and CL dependent betaine transporter that is regulated by hypertonicity. *J. Biol. Chem.*, 267, 649-652

Yancey, P.H., Clark, M.E., Hand, S.C., Bowlus, R.D., Somero, G.N. (1993) Living with water stress; evolution of osmolyte systems. *Science* 217,1214–1222.

Zhang, N., Wei, W., Mody, I., Houser, C.R. (2007) Altered localization of GABA-A receptor subunits on dentate granule cell dendrites influences tonic and phasic inhibition in a mouse model of epilepsy. *J. Neurosci.* 27, 7520–7531

Zhu, X. M. and Ong, W.Y. (2004) Changes in GABA transporters in the rat hippocampus after ainate-Induced neuronal Injury: decrease in GAT-1 and GAT-3 but upregulation of betaine/GABA transporter BGT-1. *J. Neurosci. Res.* 77, 402–409

In: Inositol
Editors: H. Rocha and M. Cardoso
ISBN: 978-1-62808-520-4

Chapter 4

The Contribution of Inositol Trisphosphate Signaling to Cardiac Pathologies

Elizabeth A. Woodcock
Molecular Cardiology Laboratory, Baker IDI Heart and Diabetes Institute, Melbourne, Victoria, Australia

Abstract

Inositol(1,4,5) trisphosphate is generated at the cell surface following stimulation and via activation of its receptors on intracellular stores, is the prime regulator of cellular Ca^{2+} responses. However, Ca^{2+} regulation in the heart depends most importantly on Ca^{2+}-induced Ca^{2+} release controlled by ryanodine receptors on the sarcoplamic reticulum, raising questions about the contribution of IP_3 and its receptors (IP_3-R). Furthermore, cardiomyocytes show only limited activity in generating IP_3 and the level of expression of IP_3-R is low. However, both IP_3 generation and IP_3-R expression are heightened in many cardiac pathologies, including hypertrophy, heart failure and in atrial fibrillation. IP_3 has been suggested to contribute to disease progression by promoting arrhythmia, hypertrophy and chamber dilatation. Experimental studies have argued both for and against this proposal. Many factors are altered in diseased myocardium and the challenge is to identify those that critically influence outcomes. In the current review, the evidence that IP_3 and IP_3-R contribute to heart disease is critically re-evaluated.

$Ins(1,4,5)P_3$ Responses

Inositol(1,4,5)*tris*phosphate ($Ins(1,4,5)P_3$) is a regulator of intracellular Ca^{2+} generated in response to stimulation by receptor activation or mechanical stress [1, 2]. $Ins(1,4,5)P_3$ is generated by phospholipase C (PLC) mediated hydrolysis of the minor plasma membrane phospholipid, phosphatidylinositol(4,5)*bis*phosphate ($PtdIns(4,5)P_2$) [3]. Cleavage of $Ins(1,4,5)P_3$ from $PtdIns(4,5)P_2$ leaves *sn*-1,2-diacylglycerol (DAG), a neutral lipid that remains in the membrane and has a complex signaling profile of its own. This $Ins(1,4,5)P_3$/DAG response follows activation of specific receptors on the cell surface, some of which belong to the seven transmembrane receptors category, also called G protein coupled receptors (GPCR), and others are growth factor receptors [4]. To decipher the role of $Ins(1,4,5)P_3$ in heart, it is important to first review the current understanding of how $Ins(1,4,5)P_3$ functions in non-excitable cells, where $Ins(1,4,5)P_3$ is the primary regulator of Ca^{2+} [5]. Activation of PLC-coupled receptors results in a rapid generation of $Ins(1,4,5)P_3$, along with DAG. $Ins(1,4,5)P_3$ is immediately metabolised by both phosphorylation and dephosphorylation pathways generating a plethora of inositol derivatives, some of which have functional roles [1]. $Ins(1,4,5)P_3$ is highly regulated and heightened generation is immediately balanced by increased metabolism, primarily by IP_3-5-phosphatase localized predominantly at the plasma membrane [6, 7]. This means that measurable changes in $Ins(1,4,5)P_3$ itself are generally rapid, quantitatively small and transient [8]. $Ins(1,4,5)P_3$ binds to high affinity receptors (IP_3-R) expressed on intracellular Ca^{2+} stores close to the plasma membrane, initiating Ca^{2+} release that perpetuates via a process of Ca^{2+}-induced Ca^{2+} release [9, 10]. The affinity of $Ins(1,4,5)P_3$ for all three classes of IP_3-R is substantially higher than for the phosphatases and kinases responsible for its metabolism, and thus $Ins(1,4,5)P_3$ can readily activate IP_3-R adjacent to the site of its generation [11, 12]. The consequent depletion of Ca^{2+} from the intracellular store activates a second process, whereby Ca^{2+} enters from the extracellular milieu to refill the stores, by store-operated Ca^{2+} entry (SOCE) [13, 14]. SOCE involves close interaction between a Ca^{2+}-sensing protein, stromal interacting molecule 1 (STIM1), and plasma membrane Ca^{2+} channels, including ORAI and canonical transient receptor potential channels (TrpC) [15-17]. Thus, the functioning of an active $Ins(1,4,5)P_3$ signaling response generally requires the presence of receptor-activated PLC, IP_3-R localized on an intracellular Ca^{2+} store close to the site of $Ins(1,4,5)P_3$ generation and a system that orchestrates signaling between the Ca^{2+} store and plasma membrane Ca^{2+} channels to re-fill the stores. Not all of these system are active in cardiomyocytes and therefore signaling by Ins(1,4,5)P_3 must follow a different program in heart tissue.

Ins(1,4,5)P_3 Responses in Heart

Ca^{2+} is central to the regular contraction/relaxation cycling of the heart that underlies its pump function. In heart, Ca^{2+} responses are initiated by sarcolemmal depolarization, followed by Ca^{2+} entry and amplification of the initiating Ca^{2+} signal orchestrated by ryanodine receptors (RyR) on the sarcoplasmic reticulum (SR) independently of Ins(1,4,5)P_3 [18]. The Ca^{2+} signals generated by the ryanodine receptors are the prime mechanism responsible for the initiation of contraction. Accordingly, receptors that are associated with PLC activation and Ins(1,4,5)P_3 generation, most notably α_1-adrenergic receptors, endothelin receptors and angiotensin II receptors (AT-1) are not acute regulators of cardiac contractility, at least under physiological conditions [19-21]. Instead, these receptor types are often associated with pathological hypertrophy and arrhythmogenesis [22].

There are a number of factors that prevent receptor-mediated PLC activation and Ins(1,4,5)P_3 generation from contributing significantly to Ca^{2+}-regulated contractile responses. First, the expression level of IP_3-R is very low, typically less than 1/50 of that of the ryanodine receptors [23] and for the most part these are not located on the sarcoplasmic reticulum (SR) close to the site of Ins(1,4,5)P_3 generation [24]. IP_3-R in heart will be discussed subsequently. Second, even though cardiomyocytes express receptors, G proteins and PLC enzymes at levels not markedly different from those in other cell types, the PLC response itself is slow and weak compared with responses in non-excitable cells [25].

Under some pathological conditions, PLC responses in heart are heightened and thus more Ins(1,4,5)P_3 is generated following receptor activation. We first reported heightened Ins(1,4,5)P_3 generation under the acute conditions of post-ischemic reperfusion [26, 27], a condition associated with arrhythmia and with contractile dysfunction. Under these pathological conditions there was an overall increase in the generation of Ins(1,4,5)P_3 and its metabolites by 5-10 fold compared with responses under normoxic conditions. Furthermore, under these acute pathological conditions, increased Ins(1,4,5)P_3 metabolism did not balance the heightened PLC activity and quantitative increases in the content of Ins(1,4,5)P_3 were observed [27, 28].

Subsequently, heightened PLC activation and Ins(1,4,5)P_3 generation were demonstrated in some chronic cardiac pathologies. Hypertrophied hearts from mice subjected to chronic pressure overload had heightened PLC activity and Ins(1,4,5)P_3 generation [29]. Heightened Ins(1,4,5)P_3 generation was also observed in dilated atria from patients with valvular heart disease as well as in atria from mice with severe dilated cardiomyopathy [30]. In contrast to Ins(1,4,5)P_3 responses during acute post-ischemic reperfusion, the increased generation of Ins(1,4,5)P_3 in chronically diseased myocardium was largely balanced by increased turnover such that significant increases in Ins(1,4,5)P_3 itself were not obvious, although there were clear increases in Ins(1,4,5)P_3 metabolites [30, 31]. The effect of acute ischemia/reperfusion or chronic heart failure on Ins(1,4,5)P_3 generation and metabolism is shown diagrammatically in Figure 1.

Increased Ins(1,4,5)P_3 turnover to balance generation does not necessarily preclude heightened IP_3-R binding, as the affinity of the IP_3-R for Ins(1,4,5)P_3 is substantially higher than for IP_3-5-phosphatase, the major pathway of Ins(1,4,5)P_3 metabolism in cardiomyocytes [32, 33]. But the findings are suggestive that Ins(1,4,5)P_3 is more likely to be a major contributor to pathology under the acute conditions of post-ischemic reperfusion rather than in chronic disease situations.

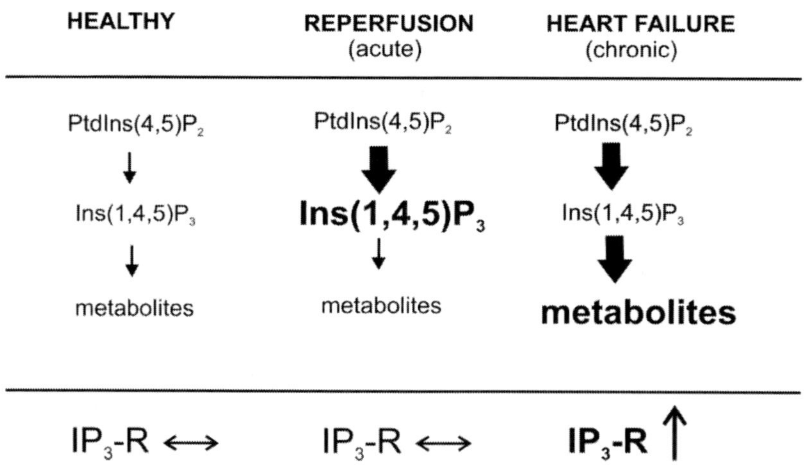

Figure 1. Diagram showing the generation and metabolism of Ins(1,4,5)P_3 under acute (ischemia/reperfusion) and chronic (heart failure) pathological conditions. Changes in expression of IP3-R are indicated in parallel.

IP_3-R Expression and Localization in Cardiomyocytes

IP_3-R are expressed in cardiomyocytes, albeit at low level as mentioned above [23]. In non-excitable cells IP_3-R are localized on endoplasmic reticulum Ca^{2+} stores close to the site of Ins(1,4,5)P_3 generation at the cell membrane [5]. Ins(1,4,5)P_3 acting on its receptors leads to depletion of the IP_3-sensitive stores and this initiates SOCE resulting in replenishment of the intracellular Ca^{2+} stores from extracellular sources [34]. Clearly a SOCE response is essential for sustained activity in non-excitable cells, but in cardiomyocytes replenishment of intracellular Ca^{2+} stores occurs by other mechanisms. Ca^{2+} enters cardiomyocytes primarily through voltage regulated L-type Ca^{2+} channels and sarcoplasmic reticulum Ca^{2+} stores are replenished by the sarco-endoplamic reticulum ATPase (SERCA) [18].

Contracting ventricular myocytes express only the type 2 IP_3-R (IP_3-R(2)). These are expressed at low level and are localized primarily on the peri-nuclear membrane [35, 36]. This raises a number of questions about the functional role of Ins(1,4,5)P_3 generated at the sarcolemma and its receptors localized at the nucleus. Atrial myocytes also express IP_3-R(2) on the peri-nuclear membrane, but in these cells a small contingent of IP_3-R(2) is also localized in the subsarcolemmal region [37]. Similarly, subsarcolemmal IP_3-R(2) are present in pacemaker cells [38]. Conducting myocytes, in contrast, express IP_3-R(1) and there is evidence of expression in the sub-sarcolemmal region [39].

The expression of IP_3-R(2) in heart is increased under pathological conditions. The first report of this was from the Marks' laboratory who described increased expression of IP_3-R along with lowered ryanodine receptor content in ventricular tissue from patients with heart failure [23]. This finding of increased IP_3-R expression was subsequently extended to rabbit and mouse models of dilated cardiomyopathy [40, 41] and pressure overload hypertrophy [41]. Increased IP_3-R expression has also been reported in isolated neonatal rat ventricular myocytes when stimulated to hypertrophy [42]. In atrial tissue, IP_3-R expression was reported to be elevated in atrial appendage from patients with atrial fibrillation compared with those in sinus rhythm [43] and one study reported a return to control expression levels after cardioversion to treat the arrhythmia [44]. Heightened expression of IP_3-R is thus a common feature of cardiac pathology and it has been argued that the increased receptors contribute to disease progression, especially in atrial fibrillation where expression changes were reversed by treating the arrhythmia [44]. However, the data are also compatible with the concept that the change in IP_3-R expression is caused by the disease and reflects a de-differentiation of the cardiomyocytes resulting from chronic heart failure. Changes in IP_3-R expression in cardiac pathologies is shown in Figure 1, along with changes in the generation of Ins(1,4,5)P_3.

Pathways of Cardiomyocyte Hypertrophy and the Possible Contribution of Ins(1,4,5)P_3

Hearts undergo hypertrophy, resulting in increased heart mass, in response to both pathological and physiological stimuli [45]. Pathological hypertrophy is a response to such insults as pressure or volume overload, myocardial infarction and genetic factors, whereas physiological hypertrophy occurs during post-natal growth and can be induced by exercise [46, 47]. Pathological hypertrophy progresses to heart failure and is associated with structural remodeling and arrhythmia [45]. Physiological hypertrophy, on the other hand, involves an increase in the size and contractile performance of the heart without causing pathological changes, and is generally regarded as protective [48]. The two broad classes of hypertrophy are mediated by different mechanisms; pathological hypertrophy involving mechanisms initiated by G protein coupled receptors and physiological hypertrophy being associated with pathways downstream of growth factor receptors [45, 46, 49, 50]. Ins(1,4,5)P_3 has mostly been considered a possible contributor to pathological hypertrophy.

Mechanisms of hypertrophy have been studied in the neonatal rat ventricular myocyte model, where activation of receptors coupled to Gq induce a growth pattern that reflects, to some extent, pathological hypertrophy [51-53]. Gq initiates its downstream responses primarily by activating PLCβ family members and would be expected to produce Ins(1,4,5)P_3 and DAG, as outlined earlier. On this basis, one or both of these factors (Ins(1,4,5)P_3 or DAG) would be expected to initiate hypertrophic signaling. Evidence for an involvement of DAG in pathological hypertrophy has been demonstrated using transgenic mice that overexpress diacylglycerol kinase ζ (DGKζ) in heart to increase DAG metabolism [54-57]. In contrast, direct evidence for a requirement for Ins(1,4,5)P_3 has proven more difficult . Cardiomyocyte hypertrophy is mediated by a complex interplay of signaling mechanisms involving phosphorylation/dephosphorylation cascades dependent to a greater or lesser extent on changes in intracellular Ca^{2+} [58]. Even though an enormous number of different pathways have been implicated, there is general agreement about the involvement of the Ca^{2+}-activated protein phosphatase, calcineurin, and possibly also the Ca^{2+}-dependent protein kinase calmodulin-dependent protein kinase IIδ [49, 58, 59]. The central role of Ca^{2+} in cardiomyocyte hypertrophy is interesting, because cardiomyocytes *in situ* undergo continual cycling of intracellular Ca^{2+} responsible for maintaining the heart beat. This raises questions about the source of the Ca^{2+} that activates hypertrophic pathways and how it is distinguished from 'systolic' (contractile) Ca^{2+}'.

Different mechanisms have been suggested to account for this Ca^{2+}, and one of the first suggested involved Ins(1,4,5)P_3. It was proposed that Ins(1,4,5)P_3 acting on its receptors localized on the SR generated quantitatively small Ca^{2+} responses that selectively activated calcineurin in the sub-sarcolemmal region of the cell [60]. This idea has largely been discounted because of the localization of IP_3-R on the nuclear membrane where any Ca^{2+} released is unlikely to activate calcineurin. This model is outlined diagrammatically in Figure 2A. In agreement with this, when type 2 IP_3-R were expressed on the sub-sarcolemmal SR a small hypertrophic response was observed, but whether this involved calcineurin was not reported [61]. Current data suggest that it is more likely that calcineurin is activated by Ca^{2+} entry through TrpC [62-64]. TrpC are activated downstream of receptors coupled to PLC activation and some subtypes, TrpC3 and TrpC6, are directly activated by DAG [65]. Recently, the suggestion has been made that a sub-group of L-type Ca^{2+} channels also contribute to this 'hypertrophic' Ca^{2+} pool [62, 66].

The localization of IP_3-R around the nucleus suggests a possible involvement of Ins(1,4,5)P_3 in regulating transcriptional changes involved in hypertrophy, rather than in calcineurin regulation. The Ca^{2+}-activated protein kinase, CaMKIIδ, is scaffolded onto peri-nuclear IP_3-R well placed to respond to the small amount of Ca^{2+} released by those receptors [36, 67]. Importantly, activated CaMKIIδ was shown to phosphorylate histone deacetylase 5 (HDAC5) promoting its exclusion from the nucleus [67]. HDAC5 represses Mef2 transcriptional activity and therefore the net result of peri-nuclear CaMKIIδ activation by this mechanism is heightened transcription of Mef2-responsive genes. This suggests the possibility that activation of sarcolemmal Gq-coupled receptors generates Ins(1,4,5)P_3 that activates peri-nuclear IP_3-R, producing a localized Ca^{2+} response that is sufficient to activate CaMKIIδ and subsequently regulate transcriptional activity [68].

The most intriguing question arising from these findings is the site of origin of the Ins(1,4,5)P_3 to activate the perinuclear receptors. There are a number of possibilities and these have been discussed in a recent editorial [69]. First, the Ins(1,4,5)P_3 could be generated at the sarcolemma and diffuse to the nucleus to activate its receptors (Figure 2B). Ins(1,4,5)P_3 is a small, highly charged molecule, which would be expected to diffuse readily in the cytosol [70]. Arguing against this is that fact that Ins(1,4,5)P_3 is rapidly dephosphorylated close to its site of generation, and furthermore increased turnover generally appears to follow heightened generation of Ins(1,4,5)P_3 [7]. These considerations suggest caution, but they do not absolutely rule out the possibility that a minor fraction of the Ins(1,4,5)P_3 generated at the sarcolemma reaches IP_3-R on the perinuclear membrane and that this is sufficient to activate CaMKIIδ.

Another possibility is that Ins(1,4,5)P_3 can be generated by receptors at the nuclear membrane localized so as to activate adjacent peri-nuclear IP_3-R (Figure 2C). The idea that an inositide response was possible in the nuclear membrane and that it was entirely separate from the plasma membrane pathway was first suggested several decades ago [71, 72]. At the time, these views were controversial and there remains no general consensus about nuclear phospholipase C signaling. The idea has been revived recently with the suggestion that certain Gq-coupled receptors, specifically α_1-adrenergic receptors and angiotensin II receptors (AT1-R), can be localized on the nuclear membrane [73, 74]. However, how such receptors are activated by extracellular agonists and whether they generate Ins(1,4,5)P_3 have not been established. α_1-adrenergic receptors and AT1-R would both be expected to activate PLCβ family members, via recruitment of Gq. Some studies reported finding PLCβ subtypes, primarily PLCβ1, on the nuclear membrane [73, 75], whereas others found that PLCβ subtypes were nuclear excluded, particularly PLCβ1 [76]. Expression of the 'b' splice variant of PLCβ1, PLCβ1b, was demonstrated to be sufficient to cause cardiomyocyte hypertrophy, but this required the PLCβ1b to be localized close to the sarcolemma and dislodging PLCβ1b from the sarcolemma inhibited hypertrophy downstream of Gq or Gq-coupled receptor agonists [77]. To add further confusion, some studies have reported the PLC substrate, PtdIns(4,5)P_2, at the peri-nuclear membrane in cardiomyocytes [78], whereas others using similar methodology found PtdIns(4,5)P_2 to be exclusively localized at the sarcolemma [7, 79].

In addition to PLCβ subtypes, another class of PLC enzymes, PLCε, has been implicated in cardiomyocyte hypertrophy, based on the finding that knockdown of PLCε reduced hypertrophic responses. PLCε, unlike PLCβ subtypes, is primarily localized around the nucleus [80]. This is intriguing because most studies agree that PtdIns(4,5)P_2 is restricted to the plasma membrane [81]. A recent study provides clarification by showing that peri-nuclear PLCε hydrolyses PtdIns(4)P, rather than PtdIns(4,5)P_2. This will generate DAG but not Ins(1,4,5)P_3. DAG generated at the nucleus is well placed to activate protein kinase D to initiate HDAC5 phosphorylation and nuclear exclusion [79]. Importantly, interfering with the nuclear localization or the activation of PLCε inhibited the development of pressure overload hypertrophy and in agreement with this, hypertrophy in this model is limited by removal of DAG [82]. In marked contrast deletion of IP_3-R(2) had no effect on any aspect of the response to pressure overload, either hypertrophic growth or the decompensation into failure [41], indicating that Ins(1,4,5)P_3, unlike DAG, is not involved in this response.

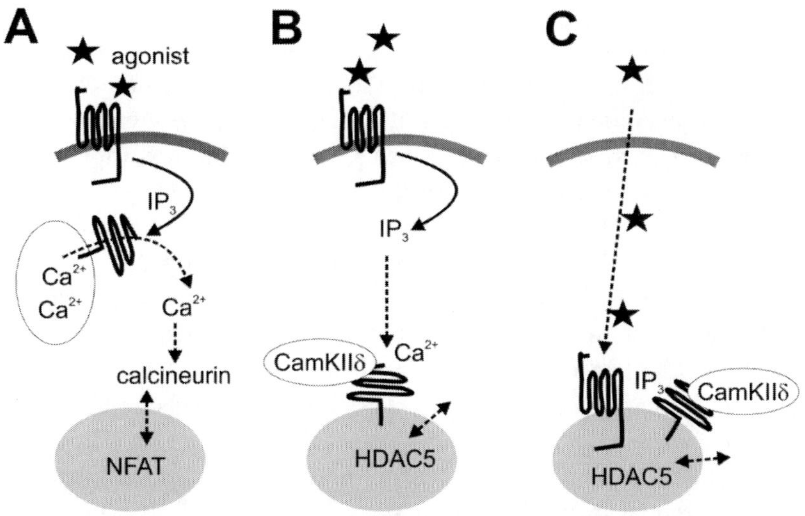

Figure 2. Different models outlining the role of Ins(1,4,5)P_3 (IP_3) in hypertrophic signaling. A. Ins(1,4,5)P_3 generated at the sarcolemma activates IP3-R on the sarcoplasmic reticulum releasing Ca^{2+} to activate calcineurin, with downstream transcriptional consequences. B. Ins(1,4,5)P_3 generated at the sarcolemma diffuses to the nucleus where it binds peri-nuclear IP3-R generating a Ca^{2+} response that activates CaMKIIδ resulting in HDAC5 phosphorylation and nuclear exclusion. C. Ins(1,4,5)P_3 is generated at the nuclear membrane where it acts on adjacent IP_3-R to generate Ca^{2+} and activate CaMKIIδ to induce HDAC nuclear export.

Most mechanistic studies of cardiac hypertrophy have involved the use of isolated cardiomyocyte cultures, where Gq is overexpressed or chronically activated by agonists promoting heightened PLCβ activation and Ins(1,4,5)P_3 generation [52]. Similarly, overexpression of Gq in heart *in vivo* causes heightened PLCβ activity and Ins(1,4,5)P_3 generation and results subsequently in hypertrophy [83]. Targeting Gq effectively limits hypertrophy in response to pressure or volume overload and this implies involvement of one of the products of PLCβ activation, Ins(1,4,5)P_3 or DAG, in pathological hypertrophy *in vivo* [84, 85, 86]. In marked contrast to the requirement for Gq in pressure overload hypertrophy, deleting IP_3-R(2) did not alter any aspect of the hypertrophic response [41].

Ins(1,4,5)P_3 in Arrhythmogenic Responses

In addition to hypertrophy, Ins(1,4,5)P_3 has been suggested to contribute to arrhythmogenesis [87-90]. As noted earlier, ischemia/reperfusion causes a rapid, transient generation of Ins(1,4,5)P_3 and this is associated with reperfusion-induced ventricular ectopic beats, tachycardia and fibrillation [26,27]. The reperfusion arrhythmias were effectively prevented by PLC inhibitors, pointing to the possibility of Ins(1,4,5)P_3-induced arrhythmias [26,91]. Ins(1,4,5)P_3 was suggested to contribute to arrhythmia by causing localized Ca^{2+} responses in the subsarcolemmal region, which initiate a Na^+/Ca^{2+} exchange current capable of initiating afterdepolarizations that can promote arrhythmia [92]. Another possibility is that Ca^{2+} generated at the SR activates CaMKIIδ scaffolded onto adjacent RyR causing diastolic Ca^{2+} leak and potentially contributing to arrhythmogenesis [93] (Figure 3). However, both of these models depends on IP_3-R being present in the subsarcolemmal region of the ventricular myocytes and, as noted earlier, ventricular IP_3-R are localized on the nuclear membrane where they are most unlikely to perturb the functioning of the Na^+/Ca^{2+} exchanger or RyR. It is difficult to imagine how perinuclear IP_3-R could contribute to acute changes in the electrophysiological properties of the myocytes and therefore it seems unlikely that Ins(1,4,5)P_3 is responsible for reperfusion arrhythmias. The effectiveness of PLC inhibitors in preventing arrhythmias under these conditions most likely reflects inhibition of DAG generation and the consequent failure of TrpC3/6 activation, or possibly regulation of PtdIns (4,5)P_2 a critical regulator of ion channels and exchangers, some of which are possible contributors to arrhythmia [94]. The other possibility is that the arrhythmias are initiated in the conducting myocytes that express IP_3-R (type 1) at higher levels than in the conducting myocardium.

IP_3-R are present on the subsarcolemmal SR of atrial myocytes, albeit at low levels, and therefore the likelihood of atrial arrhythmias initiated by Ins(1,4,5)P_3 is somewhat higher than in ventricle. IP_3-R expression is elevated in atrial tissue from humans and mice with atrial fibrillation [41, 43, 95, 96]. In support of this idea, Ca^{2+} responses to endothelin-1 in atrial myocytes were shown to be altered in IP_3-R(2)-/- mice, whereas there was no such effect in ventricle [97]. The data pointed to a small contribution of IP_3-R(2) to Ca^{2+} responses in atria that might contribute to arrhythmogenesis when PLC is activated. This idea was tested using a mouse model of dilated cardiomyopathy that shows conduction block and has a hightened incidence of atrial fibrillation, along with heightened generation of Ins(1,4,5)P_3 and increased expression of IP_3-R(2) [41, 95]. Deletion of IP_3-R(2) did not alter any of the disease parameters in this model strongly arguing against Ins(1,4,5)P_3 and its receptors as contributors to electrophysiological changes in atria or ventricles in this model [41].

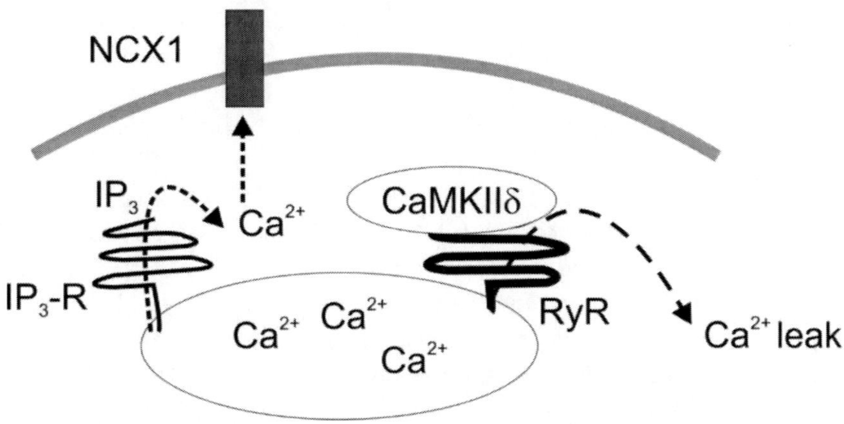

Figure 3. Potential arrhythmogenic mechanisms downstream of Ins(1,4,5)P_3 generation. Ins(1,4,5)P_3 generated at the sarcolemma activates IP3-R on the sarcocplasmic reticulum to release Ca^{2+}. This Ca^{2+} can perturb the functioning of the Na+/Ca^{2+} exchanger (NCX1) or could activate CaMKIId scaffolded onto RyR. Phosphorylation of RyR by CaMKIId results in diastolic Ca^{2+} leak that has been linked to arrhythmia.

Why Is the Question of Ins(1,4,5)P_3 Responses in Heart So Confused?

The arguments presented show that there is currently no general consensus about possible contributions of Ins(1,4,5)P_3 to cardiac function under physiological or pathological conditions. There are multiple of reasons for this. There is a general lack of specific tools to investigate Ins(1,4,5)P_3 signaling. PLC inhibitors are notoriously non-specific [98-102], and in any case PLC generates DAG as well as Ins(1,4,5)P_3, and as noted above DAG can initiate Ca^{2+} responses by activating TrpC [103]. There are no inhibitors of Ins(1,4,5)P_3 metabolism, sufficiently specific to be useful, and even if inhibitors were available they might induce artefacts by permitting responses to Ins(1,4,5)P_3 that would normally be prevented by rapid metabolism. Similarly, IP_3-R antagonists mostly also interact with other channels, particularly TrpC [104, 105]. Given that DAG, generated in parallel with Ins(1,4,5)P_3, can activate TrpC, to facilitate Ca^{2+} entry, it is obviously difficult to distinguish responses that are exclusively mediated by Ins(1,4,5)P_3 and IP_3-R. Increasing the metabolism of Ins(1,4,5)P_3 by expressing IP_3-5-phosphatase has been reported to limit hypertrophy in some studies [42], but not in others [7] and in any case, increasing Ins(1,4,5)P_3 metabolism by this method results in increased expression and activity of PLCβ1 [7] and thus more Ins(1,4,5)P_3 generation. Some studies have expressed the Ins(1,4,5)P_3-binding region of the IP_3-R to sequester Ins(1,4,5)P_3. This has been reported to inhibit hypertrophy in response to some stimuli [42, 61], but surprisingly not hypertrophy in response to pressure overload [61], which depends on Gq [85, 86] and therefore presumably on PLC activation. The Ins(1,4,5)P_3-binding region of the IP_3-R binds factors other than Ins(1,4,5)P_3 [106] and has been shown to influence autophagic responses [107]. Surprisingly, one study reported that the Ins(1,4,5)P_3-binding domain of the IP_3-R inhibited hypertrophy only when it was expressed in the nucleus [108]. This disagrees with the above reports [42,61] and also is not in keeping with the notion that Ins(1,4,5)P_3 generated at the sarcolemma diffuses through the cytosol and activates peri-nuclear IP_3-R [24] (Figure 2B). These considerations suggest that the specificity of the anti-hypertrophic activity of the Ins(1,4,5)P_3-binding domain of the IP_3-R needs to be thoroughly evaluated. A different approach was taken in a study where IP_3-R(2) were overexpressed in heart. Unlike the endogenous receptors, the exogenous IP_3-R(2) localized primarily to the SR and increased Ca^{2+} in the subsarcolemmal region [61]. Despite this, only very limited hypertrophy was reported. Given that the IP_3-R(2) overexpressing mice had severe arrhythmia there is a possibility that the hypertrophy is secondary to other pathological changes [61]. In any case, these findings are not helpful in defining the role of IP_3-R since the endogenous IP_3-R are not localized on the SR [24]. Even though mice in which the IP_3-R genes are deleted have been available for several years [7, 97], they have been used only recently to evaluate contributions of Ins(1,4,5)P_3 to cardiac pathology. Studies with the IP_3-R(2) knockout mice showed that Ins(1,4,5)P_3 did not contribute to any aspect of dilated cardiomyopathy or pressure overload hypertrophy and consequent heart failure [41]. However, this does not preclude an involvement of Ins(1,4,5)P_3 in other types of heart disease.

Conclusion

Cardiac pathologies are often associated with both increased generation of Ins(1,4,5)P_3 and increased expression of its receptors, IP_3-R. Despite this, evaluating the contribution of Ins(1,4,5)P_3 to cardiac pathology or pathophysiology has proven difficult. Ins(1,4,5)P_3 has been suggested to be involved in excitation/contraction coupling [90], pacemaker activity [38], hypertrophy [109] and arrhythmogenesis [22, 110, 111]. There is a considerable lack of agreement between the various studies suggesting caution in ascribing particular cardiac responses to Ins(1,4,5)P_3 and its receptors.

References

[1] Irvine, R. F. & Schell, M. J. (2001). Back in the water: The return of the inositol phosphates. *Nat Rev Mol Cell Biol. 2*, 327-338

[2] von Harsdorf, R., Lang, R., Fullerton, M. & Woodcock, E. A. (1989). Myocardial stretch stimulates phosphatidylinositol turnover. *Circ Res. 65*, 494-501.

[3] Divecha, N. & Irvine, R. F. (1995). Phospholipid signaling. *Cell.* 80, 269-278

[4] Irvine, RF. 20 years of Ins(1,4,5)P_3, and 40 years before. *Nat Rev Mol Cell Biol. 4*, 586-590.

[5] Streb, H., Bayerdorffer, E., Haase, W., Irvine, R. F. & Schulz, I. (1984). Effect of inositol-1,4,5-trisphosphate on isolated subcellular fractions of rat pancreas. *J Memb Biol. 81*, 241-253.

[6] Majerus, P. W., Kisseleva, M. V. & Norris, F. A. (1999).The role of phosphatases in inositol signaling reactions. *J Biol Chem. 274*, 10669-10672.

[7] Vasilevski, O., Grubb, D. R., Filtz, T. M., Yang, S., McLeod-Dryden, T. J., Luo, J., Karna, D., Chen, J. & Woodcock, E. A. (2008). Ins(1,4,5)P_3 regulates phospholipase Cβ1 expression in cardiomyocytes. *J Mol Cell Cardiol. 45*, 679-684.

[8] Woodcock, E. A., Little, P. J. & Tanner, JK. (1990). Inositol phosphate release and steroidogenesis in rat adrenal glomerulosa cells. Comparison of the effects of endothelin, angiotensin ii and vasopressin. *Biochem J. 271*, 791-796.

[9] van, Rossum, D. B., Patterson, R. L., Kiselyov, K., Boehning, D., Barrow, R. K., Gill, D. L. & Snyder, S. H. (2004). Agonist-induced Ca^{2+} entry determined by inositol 1,4,5-trisphosphate recognition. *Proc. Natl. Acad. Sci. U. S. A. 101*, 2323-2327.

[10] Dawson, A. (2003). IP_3 receptors. *Curr. Biol. 13*, R424-R424.

[11] Yoshikawa, F., Morita, M., Monkawa, T., Michikawa, T., Furuichi, T. & Mikoshiba, K. (1996). Mutational analysis of the ligand binding site of the inositol 1,4,5-trisphosphate receptor. *J Biol Chem. 271*, 18277-18284.

[12] Majerus, P. W. & York, J. D. (2009). Phosphoinositide phosphatases and disease. *J Lipid Res. 50*, S249-S254.

[13] Ma, H. T., Patterson, R. L., van, Rossum, D. B., Birnbaumer, L., Mikoshiba, K. & Gill, D. L. (2000). Requirement of the inositol trisphosphate receptor for activation of store-operated Ca^{2+} channels. *Science. 287*, 1647-1651.

[14] Putney, J. W. (2009). Capacitative calcium entry: From concept to molecules. *Immunol Rev. 231*, 10-22.

[15] Luo, X., Hojayev, B., Jiang, N., Wang, Z. V., Tandan, S., Rakalin, A., Rothermel, B. A., Gillette, T. G. & Hill, J. A. (2012). Stim1-dependent store-operated Ca^{2+} entry is required for pathological cardiac hypertrophy. *J. Mol. Cell. Cardiol. 52*, 136-147.

[16] Dingsdale, H., Voronina, S., Haynes, L., Tepikin, A. & Lur G. (2012). Cellular geography of IP_3 receptors, Stim and Orai: A lesson from secretory epithelial cells. *Biochem. Soc. Trans. 40*, 108-111.

[17] Calloway, N., Owens, T., Corwith, K., Rodgers, W., Holowka, D. & Baird, B. (2011).Stimulated association of Stim1 and Orai1 is regulated by the balance of PtdIns(4,5)P_2 between distinct membrane pools. *J. Cell Sci. 124*, 2602-2610.

[18] Bers, D. M. (2002).Cardiac excitation-contraction coupling. *Nature. 415*, 198-205.

[19] Kuraja, I. J., Tanner, J. K. & Woodcock, E. A. (1990). Endothelin stimulates phosphatidylinositol turnover in rat right and left atria. *Eur J Pharmacol. 189*, 299-306.

[20] Woodcock, E. A. & Lambert, K. A. (1995). The inositol phosphate response to thrombin in rat right atria differs from the response to noradrenaline. *Eur J Pharmacol-Molec Pharm. 291*, 213-216.

[21] Morris, J. B., Pham, T. M., Kenney, B., Sheppard, K. E. & Woodcock, E. A. (2004). UTP transactivates epidermal growth factor receptors and promotes cardiomyocyte hypertrophy despite inhibiting transcription of the hypertrophic marker gene, atrial natriuretic peptide. *J Biol Chem. 279*, 8740-8746.

[22] Roderick, H. L. & Bootman, M. D. (2007). Pacemaking, arrhythmias, inotropy and hypertrophy: The many possible facets of IP_3 signalling in cardiac myocytes. *J. Physiol.-London. 581*, 883-884.

[23] Go, L. O., Moschella, M. C., Watras, J., Handa, K. K., Fyfe, B. S. & Marks, A. R. (1995). Differential regulation of two types of intracellular calcium release channels during end-stage heart failure. *J Clin Invest. 95*, 888-894.

[24] Wu, X., Zhang, T., Bossuyt, J., Li, X. D., McKinsey, T. A., Dedman, J. R., Olson, E. N., Chen, J., Brown, J. H. & Bers, D. M. (2006). Local $InsP_3$-dependent perinuclear Ca^{2+} signaling in cardiac myocyte excitation-transcription coupling. *J Clin Invest. 116*, 675-682.

[25] Woodcock, E. A., White, L. B. S., Smith, A. I. & McLeod, J. K. (1987). Stimulation of phosphatidylinositol metabolism in the isolated, perfused rat heart. *Circ Res. 61*, 625-631.

[26] Jacobsen, A. N., Du, X. J., Lambert, K. A., Dart, A. M. & Woodcock, E. A. (1996). Arrhythmogenic action of thrombin during myocardial reperfusion via release of inositol 1,4,5-triphosphate. *Circulation. 93*, 23-26.

[27] Anderson, K., Dart, A. & Woodcock, E. (1995).Inositol phosphate release and metabolism during myocardial ischemia and reperfusion in rat heart. *Circ Res. 76*, 261-268.

[28] Harrison, S. N., Autelitano, D. J., Wang, B. H., Milano, C., Du, X. J., & Woodcock, E. A. (1998). Reduced reperfusion-induced Ins(1,4,5)P_3 generation and arrhythmias in hearts expressing constitutively active α_{1b}-adrenergic receptors. *Circ.Res. 83*, 1232-1240.

[29] Wang, B. H., Du, X. J., Autelitano, D. J., Milano, C. A. & Woodcock, E. A. (2000). Adverse effects of constitutively active α_{1b}-adrenergic receptors after pressure overload in mouse hearts. *Am J Physiol. 279*, H1079-H1086.

[30] Woodcock, E. A., Grubb, D. R., Filtz, T. M., Marasco, S., Luo, J., McLeod-Dryden, T. J., Kaye, D. M., Sadoshima, J., Du, X. J., Wong, C., McMullen, J. R. & Dart, A. M. (2009). Selective activation of the "b" splice variant of phospholipase Cβ1 in chronically dilated human and mouse atria. *J Mol Cell Cardiol. 47*, 676-683.

[31] Cooley, N., Cowley, M. J., Lin, R. C. Y., Marasco, S., Wong, C., Kaye, D. M., Dart, A. M. & Woodcock, E. A. (2012). Influence of atrial fibrillation on microRNA expression profiles in left and right atria from patients with valvular heart disease. *Physiol Genomics. 44*, 211-219.

[32] Woodcock, E., Suss, M. & Anderson, K. (1995). Inositol phosphate release and metabolism in rat left atria. *Circ Res. 76*, 252-260.

[33] Matkovich, S. J. & Woodcock, E. A. (2000). Ca^{2+}-activated but not G protein-mediated inositol phosphate responses in rat neonatal cardiomyocytes involve inositol 1,4,5-*tris*phosphate generation. *J Biol Chem. 275*, 10845-10850.

[34] Putney, J. W. (1990).Capacitative calcium entry revisited. *Cell Calcium. 11*, 611-624.

[35] Hund, T. J., Ziman, A. P., Lederer, W. J. & Mohler, P. J. (2008). The cardiac IP_3 receptor: Uncovering the role of "the other" calcium-release channel. *J Mol Cell Cardiol. 45*, 159-161.

[36] Bare, D. J., Kettlun, C. S., Liang, M., Bers, D. M. & Mignery, G. A. (2005). Cardiac type 2 inositol 1,4,5-*tris*phosphate receptor: Interaction and modulation by calcium/calmodulin-dependent protein kinase ii. *J Biol Chem. 280*, 15912-15920.

[37] Kim, J. C., Son, M. J., Subedi, K. P., Li, Y., Ahn, J. R., Woo, S. H. (2010). Atrial local Ca^{2+} signaling and inositol 1,4,5-*tris*phosphate receptors. *Prog Biophys Mol Biol. 103*, 59-70.

[38] Ju, Y. K., Liu, J., Lee, B. H., Lai, D., Woodcock, E. A., Lei, M., Cannell, M. B. & Allen, D. G. (2011). Distribution and functional role of inositol 1,4,5-*tris*phosphate receptors in mouse sinoatrial node. *Circ Res. 109*, 848-857.

[39] Stuyvers, B. D., Dun, W., Matkovich, S., Sorrentino, V., Boyden, P. A., ter, Keurs, H. E. (2005). Ca^{2+} sparks and waves in canine purkinje cells: A triple layered system of Ca^{2+} activation. *Circ Res. 97*, 35-43.

[40] Shannon, T. R., Pogwizd, S. M. & Bers, D. M. Elevated sarcoplasmic reticulum Ca^{2+} leak in intact ventricular myocytes from rabbits in heart failure. *Circ Res. 93*, 592-594.

[41] Cooley, N., Ouyang, K., McMullen, J. R., Kiriazis, H., Sheikh, F., Wu, W., Mu, Y., Du, X. J., Chen, J., Woodcock, E. A. (2013). No contribution of IP_3-R(2) to disease phenotype in models of dilated cardiomyopathy or pressure overload hypertrophy. *Circ. Heart Failure.* 6, 318-325.

[42] Higazi, D. R., Fearnley, C. J., Drawnel, F. M., Talasila, A., Corps, E. M., Ritter, O., McDonald, F., Mikoshiba, K., Bootman, M. D., Roderick, H. L. (2009). Endothelin-1-stimulated $InsP_3$-induced Ca^{2+} release is a nexus for hypertrophic signaling in cardiac myocytes. *Mol Cell. 33*, 472-482.

[43] Yamda, J., Ohkusa, T., Nao, T., Ueyama, T., Yano, M., Kobayashi, S., Hamano, K., Esato, K., Matsuzaki, M. (2001). Up-regulation of inositol 1,4,5 *tris*phosphate receptor expression in atrial tissue in patients with chronic atrial fibrillation. *J Am Coll Cardiol. 37*, 1111-1119.

[44] Gaborit, N., Steenman, M., Lamirault, G., Le, Meur, N., Le, Bouter, S., Lande, G., Leger, J., Charpentier, F., Christ, T., Dobrev, D., Escande, D., Nattel, S., Demolombe, S. (2005). Human atrial ion channel and transporter subunit gene-expression remodeling associated with valvular heart disease and atrial fibrillation. *Circulation. 112*, 471-481.

[45] Liao, R. L. & Force, T. Not all hypertrophy is created equal. (2007). *Circ.Res. 101*, 1069-1072.

[46] Maillet, M., van, Berlo, J. H. & Molkentin, J. D. (2013). Molecular basis of physiological heart growth: Fundamental concepts and new players. *Nat. Rev. Mol. Cell Biol. 14*, 38-48.

[47] Wilkins, B. J., Dai, Y. S., Bueno, O. F., Parsons, S. A., Xu, J., Plank, D. M., Jones, F., Kimball, T. R. & Molkentin, J. D. (2004). Calcineurin/NFAT coupling participates in pathological, but not physiological, cardiac hypertrophy. *Circ.Res. 94*, 110-118.

[48] McMullen, J. R., Amirahmadi, F., Woodcock, E. A., Schinke-Braun, M., Bouwman, R. D., Hewitt, K. A., Mollica, J. P., Zhang, L., Zhang, Y. Y., Shioi, T., Buerger, A., Izumo, S., Jay, P. Y. & Jennings, G. L. (2007). Protective effects of exercise and phosphoinositide 3-kinase(p110 alpha) signaling in dilated and hypertrophic cardiomyopathy. *Proc. Natl. Acad. Sci. U. S. A. 104*, 612-617.

[49] Anderson, M. E., Brown, J. H. & Bers, D. M. (2011). CaMKII in myocardial hypertrophy and heart failure. *J Mol Cell Cardiol. 51*, 468-473.

[50] Tardiff, J. C. (2006). Cardiac hypertrophy: Stressing out the heart. *J Clin Invest. 116*, 1467-1470.

[51] McKinsey, T. A. & Olson, E. N. (1999). Cardiac hypertrophy: Sorting out the circuitry. *Curr Opin Genet Dev. 9*, 267-274.

[52] Dorn, G. W. 2nd. & Brown, J. H. (1999). Gq signaling in cardiac adaptation and maladaptation. *Trends Cardiovasc Med. 9*, 26-34.

[53] Schaub, M. C., Hefti, M. A., Harder, B. A. & Eppenberger, H. M. (1997). Various hypertrophic stimuli induce distinct phenotypes in cardiomyocytes. *J Molr Med Jmm. 75*, 901-920.

[54] Hirose, M., Takeishi, Y., Niizeki, T., Shimojo, H., Nakada, T., Kubota, I., Nakayama, J., Mende, U. & Yamada, M. (2009). Diacylglycerol kinase ζ inhibits Gαq-induced atrial remodeling in transgenic mice. *Heart Rhythm, 6*, 78-84.

[55] Niizeki, T., Takeishi, Y., Kitahara, T., Arimoto, T., Koyama, Y., Goto, K., Mende, U. & Kubota, I. (2008). Diacylglycerol kinase ζa rescues Gaq-induced heart failure in transgenic mice. *Circ J. 72*, 309-317.

[56] Niizeki, T., Takeishi, Y., Arimoto, T., Takahashi, H., Shishido, T., Koyama, Y., Goto, K., Walsh, R. A. & Kubota, I. (2007). Cardiac-specific overexpression of diacylglycerol kinase ζ attenuates left ventricular remodeling and improves survival after myocardial infarction. *Am J Physiol, 292*, H1105-1112.

[57] Arimoto, T., Takeishi, Y., Takahashi, H., Shishido, T., Niizeki, T., Koyama, Y., Shiga, R., Nozaki, N., Nakajima, O., Nishimaru, K., Abe, J., Endoh, M., Walsh, R. A., Goto, K. & Kubota, I. (2006). Cardiac-specific overexpression of diacylglycerol kinase ζ prevents Qq protein-coupled receptor agonist-induced cardiac hypertrophy in transgenic mice. *Circulation. 113*, 60-66.

[58] Molkentin, J. D. (2000). Calcineurin and beyond : Cardiac hypertrophic signaling. *Circ.Res. 87*, 731-738.

[59] Molkentin, J. D., Lu, J. R., Antos, C. L., Markham, B., Richardson, J., Robbins, J., Grant, S. R. & Olson, E. N. (1998). A calcineurin-dependent transcriptional pathway for cardiac hypertrophy. *Cell. 93*, 215-228.

[60] Goonasekera, S. A. & Molkentin, J. D. (2012). Unraveling the secrets of a double life: Contractile versus signaling Ca^{2+} in a cardiac myocyte. *J Mol Cell Cardiol. 52*, 317-322

[61] Nakayama, H., Bodi, I., Maillet, M., DeSantiago, J., Domeier, T. L., Mikoshiba, K., Lorenz, J. N., Blatter, L. A., Bers, D. M. & Molkentin, J. D. (2010). The IP_3 receptor regulates cardiac hypertrophy in response to select stimuli. *Circ Res. 107*, 659-666.

[62] Gao, H., Wang, F., Wang, W., Makarewich, C. A., Zhang, H. Y., Kubo, H., Berretta, R. M., Barr, L. A., Molkentin, J. D. & Houser, S. R. (2012). Ca^{2+} influx through l-type Ca^{2+} channels and transient receptor potential channels activates pathological hypertrophy signaling. *J. Mol. Cell. Cardiol. 53*, 657-667

[63] Eder, P. & Molkentin, J. D. (2011). Trpc channels as effectors of cardiac hypertrophy. *Circ Res. 108*, 265-272.

[64] Wu, X., Eder, P., Chang, B. & Molkentin, J. D. (2010). TrpC channels are necessary mediators of pathologic cardiac hypertrophy. *Proc Natl Acad Sci U S A. 107*, 7000-7005.

[65] Lemonnier, L., Trebak, M. & Putney, J. W. (2008). Complex regulation of the TrpC3, 6 and 7 channel subfamily by diacylglycerol and phosphatidylinositol-4,5-*bis*phosphate. *Cell Calcium. 43*, 506-514.

[66] Makarewich, C. A., Correll, R. N., Gao, H., Zhang, H., Yang, B., Berretta, R. M., Rizzo, V., Molkentin, J. D. & Houser, S. R. (2012). A caveolae-targeted l-type Ca^{2+} channel antagonist inhibits hypertrophic signaling without reducing cardiac contractility. *Circ Res. 110*, 669-674

[67] Bossuyt, J., Helmstadter, K., Wu, X., Clements-Jewery, H., Haworth, R. S., Avkiran, M., Martin, J. L., Pogwizd, S. M. & Bers, D. M. (2008). Ca^{2+}/calmodulin-dependent protein kinase IIδ and protein kinase D overexpression reinforce the histone deacetylase 5 redistribution in heart failure. *Circ Res. 102*, 695-702.

[68] Mishra, S., Gray, C. B. B., Miyamoto, S., Bers, D. M. & Brown, J. H. (2011). Location matters clarifying the concept of nuclear and cytosolic camkii subtypes. *Circ.Res. 109*, 1354-NIL_1104.

[69] Bers, D. M. (2013). Membrane receptor neighborhoods: Snuggling up to the nucleus. *Circ Res. 112*, 224-226.

[70] Allbritton, N. L., Meyer, T. & Stryer, L. (1992). Range of messenger action of calcium ion and inositol 1,4,5-*tris*phosphate. *Science. 258*, 1812-1815

[71] Divecha, N., Banfic, H. & Irvine, R. F. (1993). Inositides and the nucleus and inositides in the nucleus. *Cell.* 74, 405-407

[72] Divecha, N., Banfic, H., Irvine, R. F. (1991). The polyphosphoinositide cycle exists in the nuclei of Swiss 3T3 cells under the control of a receptor (for IGF-I) in the plasma membrane, and stimulation of the

cycle increases nuclear diacylglycerol and apparently induces translocation of protein kinase-C to the nucleus. *EMBO J. 10*, 3207-3214.

[73] Wright, C. D., Chen, Q. H., Baye, N. L., Huang, Y., Healy, C. L., Kasinathan, S. & O'Connell, T. D. (2008). Nuclear α_1-adrenergic receptors signal activated erk localization to caveolae in adult cardiac myocytes. *Circ.Res. 103*, 992-1000.

[74] Tadevosyan, A., Maguy, A., Villeneuve, L. R., Babin, J., Bonnefoy, A. Allen, B. G. & Nattel, S. (2010). Nuclear-delimited angiotensin receptor-mediated signaling regulates cardiomyocyte gene expression. *J Biol Chem. 285*, 22338-22349.

[75] Faenza, I., Bregoli, L., Ramazzotti, G., Gaboardi, G., Follo, M. Y., Mongiorgi, S., Billi, A. M., Manzoli, L., Martelli, A. M. & Cocco, L. Nuclear phospholipase C β1 and cellular differentiation. *Front. Biosci.* 2008).13, 2452-2463.

[76] Grubb, D. R., Vasilevski, O., Huynh, H. & Woodcock, E. A. (2008). The extreme c-terminal region of phospholipase Cβ1 determines subcellular localization and function). the "b" splice variant mediates alpha1-adrenergic receptor responses in cardiomyocytes. *FASEB J. 22*, 2768-2774.

[77] Filtz, T. M., Grubb, D. R., McLeod-Dryden, T. J., Luo, J. & Woodcock, E. A. (2009). Gq-initiated cardiomyocyte hypertrophy is mediated by phospholipase Cβ1b. *FASEB J. 23*, 3564-3570.

[78] Ibarra, C., Vicencio, J. M., Estrada, M., Lin, Y. B., Rocco, P., Rebellato, P., Munoz, J. P., Garcia-Prieto, J., Quest, A. F. G., Chiong, M., Davidson, S. M., Bulatovic, I., Grinnemo, K. H., Larsson, O., Szabadkai, G., Uhlen, P., Jaimovich, E. & Lavandero, S. (2013). Local control of nuclear calcium signaling in cardiac myocytes by perinuclear microdomains of sarcolemmal insulin-like growth factor 1 receptors. *Circ.Res. 112*, 1270-NIL_1278.

[79] Zhang, L., Malik, S., Pang, J., Wang, H., Park, K. M., Yule, D. I., Blaxall, B. C. & Smrcka, A. V. (2013). Phospholipase Cε hydrolyzes perinuclear phosphatidylinositol 4-phosphate to regulate cardiac hypertrophy. *Cell. 153*, 216-227.

[80] Zhang, L., Malik, S., Kelley, G. G., Kapiloff, M. S. & Smrcka, A. V. (2011). Phospholipase Cε scaffolds to muscle-specific a kinase anchoring protein (MAKAPβ) and integrates multiple hypertrophic stimuli in cardiac myocytes. *J Biol Chem. 286*, 23012-23021.

[81] Falkenburger, B. H., Jensen, J. B., Dickson, E. J., Suh, B. C. & Hille, B. (2010). Phosphoinositides: Lipid regulators of membrane proteins. *J. Physiol.-London*. *588*, 3179-3185.

[82] Harada, M., Takeishi, Y., Arimoto, T., Niizeki, T., Kitahara, T., Goto, K., Walsh, R. A. & Kubota, I. (2007). Diacylglycerol kinase ζ attenuates pressure overload-induced cardiac hypertrophy. *Circ J*. *71*, 276-282.

[83] Sakata, Y., Hoit, B. D., Liggett, S. B., Walsh, R. A. & Dorn, G. W. (1998). 2nd. Decompensation of pressure-overload hypertrophy in Gαq-overexpressing mice. *Circulation*. *97*, 1488-1495.

[84] Esposito, G., Rapacciuolo, A., Naga, Prasad, S. V., Takaoka, H., Thomas, S. A., Koch, W. J. & Rockman, H. A. (2002). Genetic alterations that inhibit *in vivo* pressure-overload hypertrophy prevent cardiac dysfunction despite increased wall stress. *Circulation*. *105*, 85-92.

[85] Akhter, S. A., Luttrell, L. M., Rockman, H. A., Iaccarino, G., Lefkowitz, R. J. & Koch, W. J. (1998). Targeting the receptor-Gq interface to inhibit *in vivo* pressure overload myocardial hypertrophy. *Science*. *280*, 574-577.

[86] Wettschureck, N., Rutten, H., Zywietz, A., Gehring, D., Wilkie, T. M., Chen, J., Chien, K. R. & Offermanns, S. (2001). Absence of pressure overload induced myocardial hypertrophy after conditional inactivation of Gαq/Gα11 in cardiomyocytes. *Nat Med*. *7*, 1236-1240.

[87] Wehrens, X. H., Lehnart, S. E. & Marks, A. R. (2005). Intracellular calcium release and cardiac disease. *Annu Rev Physiol*. *67*, 69-98.

[88] Shilkrut, M., Gealekman, O., Rosen, D., Berke, G., Woodcock, E. & Binah, O. (2001). Electrophysiologic perturbations and arrhythmogenic activity caused by activation of the Fas receptor in murine ventricular myocytes: Role of the inositol trisphosphate pathway. *J Cardiovasc Electrophysiol*. *12*, 185-195..

[89] Woodcock, E. A., Arthur, J. F. & Matkovich, S. J. (2000). Inositol 1,4,5-*tris*phosphate and reperfusion arrhythmias. *Clin Exp Pharmacol Physiol*. *27*, 734-737.

[90] Lipp, P., Laine, M., Tovey, S. C., Burrell, K. M., Berridge, M. J., Li, W. & Bootman, M. D. (2000). Functional $InsP_3$ receptors that may modulate excitation-contraction coupling in the heart. *Curr Biol*. *10*, 939-942..

[91] Du, X-J., Anderson, K., Jacobsen, A., Woodcock, E. & Dart, A. (1995). Suppression of ventricular arrhythmias during ischaemia-reperfusion by agents inhibiting $Ins(1,4,5)P_3$ release. *Circulation*. *91*, 2712-2716.

[92] Van, Wagoner, D. R. & Bond, M. (2001). Reperfusion arrhythmias: New insights into the role of the Na+/ Ca^{2+} exchanger. *J Mol Cell Cardiol. 33*, 2071-2074.
[93] Sossalla, S., Fluschnik, N., Schotola, H., Ort, K. R., Neef, S., Schulte, T., Wittkopper, K., Renner, A., Schmitto, J. D., Gummert, J., El-Armouche, A., Hasenfuss, G. & Maier, L. S. (2010). Inhibition of elevated Ca^{2+}/calmodulin-dependent protein kinase II improves contractility in human failing myocardium. *Circ.Res. 107*, 1150-U1215.
[94] Woodcock, E. A., Kistler, P. M. & Ju, Y. K. (2009). Phosphoinositide signalling and cardiac arrhythmias. *Cardiovasc Res. 82*, 286-295.
[95] Pretorius, L., Du, X. J., Woodcock, E. A., Kiriazis, H., Lin, R. C., Marasco, S., Medcalf, R. L., Ming, Z., Head, G. A., Tan, J. W., Cemerlang, N., Sadoshima, J., Shioi, T., Izumo, S., Lukoshkova, E. V., Dart, A. M., Jennings, G. L. & McMullen, J. R. (2009). Reduced phosphoinositide 3-kinase (p110α) activation increases the susceptibility to atrial fibrillation. *Am J Pathol. 175*, 998-1009.
[96] Lamirault, G., Gaborit, N., Le, Meur, N., Chevalier, C., Lande, G., Demolombe, S., Escande, D., Nattel, S., Leger, J. J. & Steenman, M. (2006). Gene expression profile associated with chronic atrial fibrillation and underlying valvular heart disease in man. *J Mol Cell Cardiol. 40*, 173-184.
[97] Li, X., Zima, A. V., Sheikh, F., Blatter, L. A. & Chen, J. (2005). Endothelin-1-induced arrhythmogenic Ca^{2+} signaling is abolished in atrial myocytes of inositol-1,4,5-*tris*phosphate (IP_3)-receptor type 2-deficient mice. *Circ Res. 96*, 1274-1281.
[98] Burgdorf, C., Schafer, U., Richardt, G. & Kurz, T. (2010). U73122, an aminosteroid phospholipase C inhibitor, is a potent inhibitor of cardiac phospholipase D by a PIP_2-dependent mechanism. *J. Cardiovasc. Pharmacol. 55*, 555-559.
[99] Klose, A., Huth, T. & Alzheimer, C. (2008). 1-[6-[[(17beta)-3-methoxyestra-1,3,5(10)-trien-17-yl]amino]hexyl]-1h-pyrro le-2,5-dione (u73122) selectively inhibits kir3 and bk channels in a phospholipase C-independent fashion. *Mol Pharmacol. 74*, 1203-1214.
[100] Muto, Y., Nagao, T. & Urushidani, T. (1997). The putative phospholipase C inhibitor U73122 and its negative control, U73343, elicit unexpected effects on the rabbit parietal cell. *J. Pharmacol. Exp. Ther. 282*, 1379-1388.
[101] Mogami, H., Mills, C. L. & Gallagher, D. V. (1997). Phospholipase C inhibitor, U73122, releases intracellular Ca^{2+}, potentiates ins(1,4,5)P_3-

mediated Ca^{2+} release and directly activates inn channels in mouse pancreatic acinar cells. *Biochem. J. 324*, 645-651.

[102] Heemskerk, J. W. M., Farndale, R. W. & Sage, S. O. (1997). Effects of U73122 and U73343 on human platelet calcium signalling and protein tyrosine phosphorylation. *BBA Mol Cell Res. 1355*, 81-88.

[103] Estacion, M., Li, S., Sinkins, W. G., Gosling, M., Bahra, P., Poll, C., Westwick, J. & Schilling, W. P. (2004). Activation of human trpc6 channels by receptor stimulation. *J Biol Chem. 279*, 22047-22056.

[104] Xi, Q., Adebiyi, A., Zhao, G. L., Chapman, K. E., Waters, C. M., Hassid, A. & Jaggar, J. H. (2008). IP_3 constricts cerebral arteries via IP_3 receptor-mediated TrpC3 channel activation and independently of sarcoplasmic reticulum Ca^{2+} release. *Circ.Res. 102*, 1118-1126.

[105] Ma, H. T., Venkatachalam, K., Rys-Sikora, K. E., He, L. P., Zheng, F. & Gill, D. L. (2003). Modification of phospholipase C-γ-induced Ca^{2+} signal generation by 2-aminoethoxydiphenyl borate. *Biochem. J. 376*, 667-676.

[106] Wong, A., Grubb, D. R., Cooley, N., Luo, J. T. & Woodcock, E. A. (2013). Regulation of autophagy in cardiomyocytes by Ins(1,4,5)P_3 and IP_3-receptors. *J. Mol. Cell. Cardiol. 54*, 19-24.

[107] Maiuri, M. C., Criollo, A. & Kroemer, G. (2010). Crosstalk between apoptosis and autophagy within the beclin 1 interactome. *Embo J. 29*, 515-516.

[108] Arantes, L. A., Aguiar, C. J., Amaya, M. J., Figueiro, N. C., Andrade, L. M., Rocha-Resende, C., Resende, R. R., Franchini, K. G., Guatimosim, S. & Leite, M. F. (2012). Nuclear inositol 1,4,5-*tris*phosphate is a necessary and conserved signal for the induction of both pathological and physiological cardiomyocyte hypertrophy. *J Mol Cell Cardiol. 53*, 475-486.

[109] Drawnel, F. M., Wachten, D., Molkentin, J. D., Maillet, M., Aronsen, J. M., Swift, F., Sjaastad, I., Liu, N., Catalucci, D., Mikoshiba, K., Hisatsune, C., Okkenhaug, H., Andrews, S. R., Bootman, M. D. & Roderick, H. L. (2012). Mutual antagonism between IP3RII and miRNA-133a regulates calcium signals and cardiac hypertrophy. *J Cell Biol. 199*, 783-798.

[110] Harzheim, D., Movassagh, M., Foo, R. S. Y., Ritter, O., Tashfeen, A., Conway, S. J., Bootman, M. D. & Roderick, H. L. (2009). Increased $InsP_3Rs$ in the junctional sarcoplasmic reticulum augment Ca^{2+} transients and arrhythmias associated with cardiac hypertrophy. *Proc. Natl. Acad. Sci. U. S. A. 106*, 11406-11411.

[111] Proven, A., Roderick, H. L., Conway, S. J., Berridge, M. J., Horton, J. K., Capper, S. J. & Bootman, M. D. (2006). Inositol 1,4,5-*tris*phosphate supports the arrhythmogenic action of endothelin-1 on ventricular cardiac myocytes. *J Cell Sci. 119*, 3363-3375.

In: Inositol
Editors: H. Rocha and M. Cardoso

ISBN: 978-1-62808-520-4

Chapter 5

The Role of Inositol 1,4,5-Triphosphate Receptor in Hepatorenal Syndrome

Dong-Lei Wang* and Pei Liu†
Department of Infectious Diseases, The First Affiliated Hospital, China Medical University, Shenyang City, Liaoning Province, P. R. China

Abstract

Many hormone and neurotransmitter receptors utilize the phosphatedylinositol signaling cascade. Phosphatidylinositol 4,5-bisphosphate is converted into two putative second messengers, inositol 1,4,5-trisphosphate (IP_3) and diacylglycerol. IP_3 regulates various kinds of cellular responses. It induces the release of Ca^{2+} from the intracellular stores by binding to the IP_3 receptor (IP_3R) on the endoplasmic reticulum. Many cell types, often nonexcitable, including glomerular mesangial cells (GMC) and renal vascular smooth muscle cells (VSMC), depend on this pathway to couple external signals to intracellular Ca^{2+} release. IP_3Rs not only triggers the release of calcium from intracellular stores but also opens plasma membrane calcium channels. Hepatorenal syndrome (HRS) is a common complication of advanced cirrhosis, characterized by renal

* Dong-Lei Wang. E-mail: wdl19851130@163.com.
† Pei Liu. E-mail: syliupei2013@163.com.cn.

failure and major disturbances in circulatory function. Renal failure is caused by intense vasoconstriction of the renal circulation.

HRS develops in approximately 55% of all patients with FHF. In order to delineate the role of inositol 1,4,5-triphosphate receptor in HRS, we investigated renal expression of type 1 inositol 1,4,5-trisphosphate receptors (IP_3R1) in mice with FHF. FHF model was induced by intravenous injection of lipopolysaccharide (LPS) in D-galactosamine (GalN)-sensitized mice via the caudal vein. GalN/LPS treatment led to a marked renal dysfunction and hyperkalaemia, of which kidney perfusion pressure was also significantly enhanced. However, it did not see any significant change in renal histopathology of rats in FHF. Simultaneously, the expression of protein and mRNA levels of renal IP_3R1 increased markedly, serum TNF-α and endothelial-1(ET-1) levels reached a maximum level. We employed an inhibitor of IP_3Rs (2-APB) in this renal failure model, and found that reduction in the glomerular filtration rate (GFR) was reversed and hyperkalaemia recovered after administration of 2-APB; whereas, 2-APB did not affect the liver injury and the expression of IP_3R1. In contrast, intravenous injection of the TNF-α monoclonal antibody into the caudal vein significantly decreased the mortality rate of GalN/LPS-treated rats from 70% to zero, and significantly improved the compromised GFR. In TNF-α monoclonal antibody treated rats, serum TNF-α, ET-1 levels and expression of IP_3R1 were decreased. We had demonstrated in a recent study that TNF-α could enhance IP_3R1 expression at mRNA and protein levels in human mesangial cells (HMCs), mediated through TNFR1/PC-PLC/PKCα-dependent and TNFR2-dependent IP_3R1 pathways. Isolated perfusion kidney experiments demonstrated that in normal rat kidney preperfused with TNF-α, stimulation with ET-1 could raise the perfusion pressure evidently and overexpress IP_3R1 protein and mRNA levels. Similar results were seen in GMCs and VSMCs incubated by TNF-α, the calcium release markedly increased after treated with ET-1. The findings in the present study demonstrated that TNF-α-dependent IP_3R1 up-regulation might be a specific target in future studies of renal dysfunction during the end-stage of liver disease.

Introduction

IP_3 is a polyatomic ion with a molecular mass of 420.10g/mol. Its empirical formula is $C_6H_{15}O_{15}P_3$. It is composed of an inositol ring with three phosphate groups bound at the 1, 4, and 5 carbon positions, and three hydroxyl groups bound at positions 2, 3, and 6. It is made by hydrolysis of phosphatidylinositol 4,5-bisphosphate (PIP_2), a phospholipid that is located in the plasma membrane, by phospholipase C (PLC).

PIP_2 is converted into two putative second messengers, inositol 1,4,5-trisphosphate (IP_3) and diacylglycerol(DAG) [1, 2].

While DAG stays inside the membrane, IP_3 is soluble and diffuses through the cell. DAG potentially activates protein kinase C and consequently causes various cellular responses [3, 4]. The other second messenger, IP_3 is a soluble molecule and is capable of diffusing through the cytoplasm to the endoplasmic reticulum (ER), or the sarcoplasmic reticulum (SR) in the case of muscle cells [5, 6], once produced by PLC. IP_3 plays a key role in the regulation of the cell function by various hormonal stimuli [7]. Once at the ER, IP_3 is able to induce the release of Ca^{2+} into the cytoplasm from the intracellular stores by binding to the $InsP_3$ receptor ($InsP_3R$) [8, 9] on the surface of ER [10, 11]. IP_3 may also activate Ca^{2+} channels on the cell membrane indirectly by increasing the intracellular Ca^{2+} concentration.

IP_3 Receptor Genes Family

The docking of IP_3 to its receptor, which is called the inositol trisphosphate receptor ($InsP_3R$), was first studied using deletion mutagenesis in the early 1990s. IP_3Rs contains four independent ligand binding sites that are cooperative with respect to calcium channel opening. In mammalian cells, there are at least three isoforms of the IP_3R derived from three distinct genes. They differ from the IP_3 receptor structure previously reported in two small variably spliced segments. One segment (SI) is located within the IP_3 binding site, whereas another segment (SII) is located near putative sites for phosphorylation and ATP binding to modulate IP_3 action on the Ca^{2+} flux [12]. Therefore, we speculate that selective use of IP_3R subtypes permits a tissue specific and developmentally specific expression of functionally distinct channels.

In the kidney, all three types of IP_3R have been found which showed distinctive localization [13]. Type 1 was detected in glomerular mesangial cells and vascular smooth muscle cells. Type 2 was expressed exclusively in intercalated cells of collecting ducts from the cortex to the inner medulla. Type 3 was expressed in the collecting duct cells of the cortex to the outer medulla. [13, 14].

Yang and colleagues proposed that different types of IP_3 receptors might have different functions in the rat kidney [15]. The Type 1 IP_3R was cloned from mouse cerebellum Purkinje cells by Furuichi et al., who concluded that the glycoprotein P400 protein is identical to the IP_3R1 [8].

Since distinctive localization of different types of IP_3R, IP_3R1 plays a major role in the regulation of $[Ca^{2+}]i$ in the glomeruli. It was recently proven in the genetically manipulated mouse [16]. IP_3Rs not only triggers the release of calcium from intracellular stores but also opens plasma membrane calcium channels. Ca^{2+} influx through the L-type channel is needed for the opening of the IP_3R, which activates a cascade of Ca^{2+} release from the ER store [17]. The Ca^{2+} ions can enter the cell across the plasma membrane, be released from intracellular stores, or there is a combination of the two [18].

Compelling evidence demonstrates that the IP_3R1 opening may play a central role in apoptosis cause by hypoxia or ischemia, cells contraction, acute renal injury and other pathophysiological events and kidney diseases [17]. For example, elevated IP_3R1 might be causally related to initiating or perpetuating diseases such as atrial fibrillation and asthma.

The Role of IP_3R1 Play in HRS

Hepatorenal syndrome (HRS) is a common complication of advanced cirrhosis, characterized by renal failure and major disturbances in circulatory function and is highly fatal with a mortality of 60%~80%. Reduction in the effective circulating blood volume and subsequent hypoperfusion of the kidneys are the basic pathogenesis of HRS [19]. The pathophysiological hallmark of HRS is intense vasoconstriction of the renal circulation [20]. Patients with HRS have an obvious reduction in renal blood flow and the glomerular filtration rate (GFR) [21, 22]. There are only functional changes and relatively few histological changes in the kidneys. Renal failure will recover when there is recovery of liver function, and in the absence of a spontaneous hepatic recovery, liver transplantation will reverse the HRS [23].

Hitherto, various theories have been proposed to explain the development of HRS in cirrhosis and severe liver injury. It is considered generally that the decrement of renal plasma flow (RPF) caused by renal vasocontraction is a key factor [24]. The renal blood flow diminution leads to the development of renal dysfunction. Multiple mechanisms are involved in renal vasoconstriction [20]. The mechanism of the vasoconstriction is incompletely understood; it may be multifactorial, involving disturbances in the development of hyperdynamic circulation, lowering of renal perfusion pressure, the activation of the sympathetic nervous system, and increased synthesis of a variety of vasoactive mediators, which renders the kidneys more susceptible to a decrease in perfusion pressure [21].

Patients with fulminant hepatic failure (FHF) are at high risk for the serious complication of hepatorenal syndrome (HRS), which is a significant cause of death [25]. Since IP_3R1 plays an important role in kidney injury, we sought to explore the role of IP_3R1 in HRS. In order to delineate the mechanism, we investigated the renal expression of IP_3R1 in murine with FHF.

Renal Failure Model Induced by GalN/LPS Treatment

FHF model was induced by intravenous injection of lipopolysaccharide (LPS) in D-galactosamine (GalN)-sensitized mice via the caudal vein. Ying Wen et al. used GalN (800 mg/kg), followed by LPS (10 mg/kg) to induce FHF mice model (20 mice at 2,6,9h) [23]. Our new unpublished research used another rat model induced by a mixture of GalN (400 mg/kg) and LPS (32 μg/kg) (12 rats at 3,6,9,12,24h) in saline administrated via the tail vein at a volume of 2 ml/kg under anesthesia. Antibodies against TNF-α (5 mg/kg) or 2-APB (an inhibitor of IP_3Rs) (2 mg/kg) were administered 30 min before GalN/LPS treatment to further explore the role of TNF-α and IP_3R. Then we examined the expression level of IP_3R1 of the kidney in murine with FHF by immunohistochemistry, Western blot and RT-PCR, histologic, serum parameters, glomerular inulin space (GIS), intracellular $[Ca^{2+}]i$ in glomeruli, and kidney perfusion pressure.

Serum Parameter

GalN/LPS treatment led to a marked increase in serum levels of ALT, BUN, Cr, K^+, TNF-α and ET-1; whereas serum Na^+ and Cl^- levels were decreased 12 hrs. after the treatment. Serum TNF-α and endothelial-1(ET-1) level reached a maximum level. The mortality of mice reached 60% 9 hours after GalN/LPS administration, and reached 70% 24 h hours later in rats, respectively. Liver histologic in both models showed massive hemorrhage and hepatic necrosis as well as infiltration of inflammatory cells [23]. In contrast, renal morphology did not show any abnormalities. The blood pressure between the group of GalN/LPS treatment and the NS control group was not significantly different. Together, these data confirmed that we successfully established the animal models of FHF.

GFR of Animals with FHF

Micro osmotic pumps filled with ~100 μl of 8% FITC-inulin was inserted into the peritoneal cavity of rats in order to measure GFR. FITC-inulin clearance was measured under steady-state (equilibrium) conditions by implanted osmotic pumps. GFR was calculated from plasma inulin levels and the known pump infusion rate [26]. GalN/LPS treatment markedly decreased GFR compared to N.S. controls.

Western Blot for IP_3R1 Protein Expression

In order to quantify the changes of IP_3R1 protein, Western blot analysis was carried out to determine IP_3R1 protein levels. Renal IP_3R1 protein was increased in the GalN/LPS-treated group and the maximal level was seen at 12h when compared to NS controls in the rat model. In the mice model the maximal effect was seen at 9h [23].

The level of IP_3R1 expression can be affected by many factors, such as various cytokines. Our observation also indicates that factors that cannot directly induce Ca^{2+} flux may also regulate $[Ca^{2+}]$ i concentration by affecting IP_3R1 expression.

Consequently, the kidney became more sensitive to agonists that stimulate IP_3-mediated $[Ca^{2+}]$ i mobilization via the IP_3R1. Overexpression of IP_3R1 may lead to modulatory influences on renal blood flow and GFR. Interaction between IP_3R1 up-regulation and agonists may enhance the contracting ability of Glomerular mesangial cells (GMC) and vascular smooth muscle cells (VSMC) to vasoconstrictors. Overexpression of IP_3R1 may be an important factor in contributing to HRS progression.

RT-PCR for the Expression of the IP_3R1 mRNA

To determine if the increased IP_3R1 protein level in GalN/LPS-exposed rats was caused by enhanced gene expression, renal mRNA levels of IP_3R1 were evaluated by quantitative RT-PCR. IP_3R1 mRNA levels were gradually increased along with the duration after GalN/LPS exposure. In the rat model it increased nearly 3 fold as early as 3 hours compared to the controls, which peaked at 9 hours, and reached approximately 10-fold.

In animals with injection of anti-TNF-α antibodies, IP_3R1 mRNA levels were strikingly reduced compared to the GalN/LPS group; whereas, 2-APB did not significantly alter the enhanced IP_3R1 gene expression. Quantitative analysis demonstrated an increase in FHF mice to 1.5 fold of control IP_3R1 mRNA expression as early as 2h FHF, and it remained increased at 9 h FHF 1.7 fold. The maximal effect was seen at 6 h, and IP_3R1 mRNA expression was increased to 1.9 fold of control values [23].

Our result, that the increment in mRNA for IP_3R1 precedes the increment in protein, is consistent with the time sequence of increased mRNA levels prior to occurrence of increased protein levels. Our finding demonstrates that up-regulation of IP_3R1 mRNA levels can be explained by the increase in transcription of the IP_3R1 gene.

Determination of Glomerular Inulin Space (GIS)

The changes in GIS reflect the glomerular size, and indirectly indicate the vascular dilation/constriction status. The GalN/LPS group markedly decreased GIS compared to N.S. controls. The constriction of these cells significantly deceased renal blood flow rate and glomerular inulin space (GIS) as indicated in our study.

Therefore, these findings enhanced our understanding that signal molecules such as ET-1 and IP_3R1 may contribute to the contraction of GMC and VSMC, which results in reduced glomerular blood flow rate, and in turn decreases GFR during the HRS development [27].

Measurements of Intracellular [Ca^{2+}]I in Glomeruli

The GalN/LPS group increased [Ca^{2+}]i robustly compared to N.S. controls. The IP_3Rs is the primary cytosolic target for the initiation of [Ca^{2+}]i. The results of GIS and glomerular [Ca^{2+}]i are in accordance with the changes of the GRF and serum parameters during the GalN/LPS exposure. The constriction of these cells significantly decreased the renal blood flow rate and glomerular inulin space (GIS) as indicated in our study.

Therefore, these findings enhanced our understanding that signal molecules such as ET-1 and IP_3 may contribute to the contraction of GMCs and VSMCs, which results in reduced glomerular blood flow rate, and in turn decreases GFR during the HRS development.

Isolated Perfused Rat Kidney

Using the isolated perfused kidney technique, rat kidneys were perfused at a constant flow. Ninety minutes after preperfusion with TNF-α in the normal rat kidney, the perfusion pressure stimulated by endothelin-1 (ET-1) was detected. TNF-α didn't modify the baseline perfusion pressure until ET-1 was added in the perfusion fluid. Perfusion pressure of the TNF-α treated group was significantly enhanced, and overexpression of the IP_3R1 protein and mRNA levels was detected.

Cells Culture

TNF-α may act on the promoter of IP_3R1 mRNA in rat glomerular afferent arterioles smooth muscle cells (RASMC) and GMC and results in the overexpression of the IP_3R1 protein that stimulates the release of intracellular Ca^{2+} in RASMC and GMC and induces contraction of them both. We found similar results in HMCs [28].

The increased protein expression of IP_3R1 may be partly due to increased synthesis of the IP_3R1 as we found an increase in mRNA levels prior to observing an increase in protein levels. The implication of the increased IP_3R1 protein expression is that IP_3R1 may provide more ligand-binding sites for IP_3, and promote IP_3-mediated [Ca^{2+}]i mobilization leading to an increase in the contraction ability of HMCs to contraction and a reduction in the renal GFP, which might be an important reason for the occurrence and development of HRS.

Expression of IP_3R1 Protein with Immunohistochemistry

Our study shows that IP_3R1 is predominant in the GMC and renal VSMC, but absent in other renal cells, which is consistent with previous studies. IP_3R1 protein expressed at low levels in 3 control groups. IP_3R1-staining was up-regulated in 6 h and 9 h FHF groups. The maximal effect was seen at 9 h. IP_3R1 expressed at low levels in control groups [23].

As it is known, contraction of renal VSMC can result in a reduction in renal blood flow, while contraction of GMC can lower the filtration fraction and glomerular capillary ultrafiltration coefficient (Kf) [29].

GMCs modify GFP via alternations in the glomerular capillary filtration surface area due to its ability to contract. IP_3Rs are the intracellular Ca^{2+} release channels that play a key role in Ca^{2+} signal in cells [27]. Cell contraction is closely connected with changes in intracellular Ca^{2+} concentration.

Both GMC and renal VSMC are sensitive to vasoconstriction. The circulating concentrations of many vasoconstrictors including ET-1 and ANG II were markedly high in HRS. ET-1 is a potent vasoconstrictor of the renal vascular bed. Vasoconstrictors promote IP_3-mediated Ca^{2+} mobilization via the IP_3Rs, followed by release of stored intracellular Ca^{2+} and Ca^{2+} entry through plasma membrane channels [27, 29]. These mediators can cause renal vasoconstriction, but more importantly they can also decrease the glomerular Kf, thus causing a decline of GFR over and above that caused by renal vasoconstriction alone. As a result, GFR was decreased obviously in HRS. Therefore, our study focused on IP_3R1 as a candidate protein that may facilitate intracellular Ca^{2+} release.

Different Influencing Factors

Intracellular calcium transporters play an important role in many signal transduction pathways of the cell involving released calcium ions as a second messenger. In the kidney, many kinds of hormones exert their action via this

IP_3 pathway. Under specific conditions, it can participate in the development of pathophysiological events in the cells. It is well known that one cell possesses several hormone receptors which induce IP_3 production and the increase in $[Ca^{2+}]i$. Angiotensin II (ANG II), vasopressin, endothelin (ET), tumor necrosis factor-α (TNF-α) and various vasoconstrictor substances could increase the $[Ca^{2+}]i$ in renal arterioles and mesangial cells. In the glomeruli from wild-type mice, the $[Ca^{2+}]i$ level, measured using fura-2AM, showed a sharp increase in response to hormonal stimuli, endothelin and angiotensin II, in the Ca^{2+}-free medium [30]. The responses of $[Ca^{2+}]i$ to ANG II and endothelin in the glomeruli were markedly attenuated in IP_3R1 knockout mice.

ANG II stimulated IP_3 production and induces a rise in cytosolic $[Ca^{2+}]i$ in vascular smooth muscle (VSM) cells via IP_3R activation and release of Ca^{2+} from the SR [31]. The Ca^{2+} signal is augmented by calcium-induced calcium release (CICR) and by cyclic adeninediphosphate ribose (cADPR), which sensitizes the ryanodine-sensitive receptor (RyR) to Ca^{2+} to further amplify CICR.

The classic G protein-coupled receptor activation that results in IP_3 generation and release of $[Ca^{2+}]i$ from the SR likely provides an initial and transient burst of $[Ca^{2+}]i$, which can activate CICR from RyR [31].

ET-1 is a powerful renal vasoconstrictor that is believed to play an important role in the pathogenesis of numerous forms of vascular disease, including kidney failure [32-34]. We determined that both ET-1 and ET-2 stimulate rapid and biphasic increases in $[Ca^{2+}]i$, whereas ET-3 stimulates a smaller monophasic increase in $[Ca^{2+}]i$. ET-1 receptors have been identified in the kidney, and reports suggest that rat preglomerular microvessels express both ET_A and ET_B receptors [35, 36]. In the renal circulation, evidence suggests the presence of both ET_A and ET_B receptors that produce vasoconstriction [35-40]. This response profile is consistent with the activation of ET_A receptors. These studies also establish that ET-1 elevates $[Ca^{2+}]i$ by stimulating the release of calcium from intracellular stores and by stimulating the influx of calcium from the extracellular medium. ET-1 evokes a biphasic elevation of $[Ca^{2+}]i$ by binding to ET_A receptors at the cell surface and activating a pertussis toxin–insensitive G-protein that in turn stimulates phospholipase C (PLC) [41]. Furthermore, these results suggest that ET-1 increases $[Ca^{2+}]i$ by stimulating both the release of intracellular calcium and the influx of calcium from the extracellular medium. Hydrolysis of phosphatidylinositol by PLC generates the second messengers, IP_3 and DAG. IP_3 is a fast-acting mediator that facilitates the first phase of the response by binding to a receptor on the SR and triggering release of Ca^{2+} from intracellular stores. The second phase of the response involves Ca^{2+} influx from the extracellular fluid [38, 42]. The sustained elevation in $[Ca^{2+}]i$ that results from Ca^{2+} influx is postulated to increase smooth muscle contractility. Previous studies demonstrate that endothelin peptides cause a rise in $[Ca^{2+}]i$ in preglomerular microvascular smooth muscle cells (MVSMC) and in cultured human mesangial cells (HMC) [28, 43, 44]. Thus, ET-1 may be a candidate substance for mesangium relevant diseases.

Tumor necrosis factor-α (TNF-α), a proinflammatory cytokine, has been implicated as a central mediator in multiple homeostatic and pathologic processes. TNF-α was first described as a protein component of serum that exhibited potent antitumor activity [45]. It is produced chiefly by activated macrophages (M1), although it can be produced by many other cell types, such as CD^{4+} lymphocytes, NK cells and neurons. The 212-amino acid transmembrane pro-form, which is active during direct cell-to-cell contact, is expressed on the plasma membrane [46].

The two major receptors for TNF-α, TNF-R1 (p 55) and TNF-R2 (p 75) are expressed on the surface of neurons both in vitro and in vivo [47]. TNF-R1 is expressed in most tissues, and can be fully activated by both the membrane-bound and soluble trimeric forms of TNF-α, whereas TNF-R2 is found only in cells of the immune system, and respond to the membrane-bound form of the TNF-α homotrimer.

Many of the pro-inflammatory actions of TNF-α are mediated by TNF-R1, but in several cases, TNF-R1 activation requires the coordinate expression of TNF-R2 that can form heterocomplexes with TNFR1. As most information regarding TNF-α signaling is derived from TNF-R1, the role of TNF-R2 is likely underestimated. TNF-R2 may also potentiate responses mediated by TNF-R1 by the increase of the local concentration of surface-associated TNF-α available to TNF-R1.

Binding of homotrimeric TNF-α to either receptor can activate three important signaling pathways: the TNF receptor-associated death domain (TRADD), and the nuclear factor κB (NFκB) and Jun N-terminal kinase (JNK) signaling cascades. Through this diverse signaling array, TNF-α can regulate many physiologically important pathways in cells.

These include the regulation of cell survival, the initiation of apoptosis, synaptic transmission, cell contraction, and the alteration of gene expression [48-51]. Signaling cascades downstream of its cellular cognate receptors, as well as the resultant transcriptional responses, have received intense interest in regards to how such signals impact cellular physiology.

Cells can also regulate diverse biological processes through the regulation of intracellular ions. Specifically, increases in cytosolic Ca^{2+} can regulate gene transcription, membrane excitability, muscle contraction, and neurotransmitter release [52].

Park et al. submit that this increase in IP_3R expression may actually be a result of a release in JNK inhibition by the NFκB pathway [53]. The present study provided the first evidence that TNF-α could increase the production of IP_3R1 by the control of the TNF-R1/PC-PLC/PKCα and TNF-R2 signaling pathway in HMCs [28]. Our results were consistent with those of Park et al. in that TNF-a increased IP_3R1 mRNA and protein expression in primary murine neurons [53].

Wang J et al. demonstrated in their studies that TNF-α could enhance the expression of the IP_3R1 protein and IP_3R1 mRNA in renal vascular smooth muscle cells (RASMC), but did not influence the half-life of IP_3R1 mRNA in RASMC treated with TNF-α. The mechanisms may be that TNF-α takes part in the change of signal transduction in RASMC.

TNF-α may act on the promoter of IP_3R1 mRNA in RASMC and results in the expression of the IP_3R1 protein that stimulates release of intracellular Ca^{2+} in RASMC and induces contraction of RASMC [54].

Furthermore, Park, K, M and colleagues had demonstrated in their study that TNF-α was shown to potentiate neuronal Ca^{2+} signaling by enhancing IP_3R1 steady-state mRNA and protein levels [55]. They reported that TNF-α enhances IP_3R-mediated Ca^{2+} signals and promoter activity through a specificity protein (SP-1)-mediated process. The SP-1 binding site that exists within the human IP_3R1 promoter is the specific binding sequence for both JNK-mediated regulation as well as transcription factor binding. These data, in aggregate, indicate that the human IP_3R1 promoter contains a key SP-1 binding element 59 base pairs (+59-bp) downstream of the putative transcription start site [53].

Effect of Anti-TNF-α or 2-APB

An accumulating body of evidence suggests that the serum level of TNF-α increases significantly in patients with severe liver injury, and it has been demonstrated that the TNF-α inhibitor, oxpentifylline, reduced the incidence of HRS from 35% to 8% [56].

In order to further prove that TNF-α is involved in HRS by enhancing IP_3Rs expression, we employed an inhibitor of IP_3Rs (2-APB) in this renal failure model, and found that reduction in GFR was reversed by administration of 2-APB; whereas, 2-APB did not affect the liver injury and serum TNF-α and ET-1 levels. At the same time, renal over-expression of IP_3R1 was reduced to a normal level. In GalN/LPS-treated rats, when TNF-α was blocked with anti-TNF-α, it was observed that not only was liver failure alleviated significantly, but GFR also improved significantly. Furthermore, the mRNA data are consistent with protein levels, and up-regulated IP_3R1 mRNA expression in GalN/LPS-exposed rats was abrogated by anti-TNF-α antibodies, but not by 2-APB. GIS in the anti-TNF-α-GalN/LPS group and the 2-APB-GalN/LPS group were improved compared to that in the GalN/LPS group. The pre-administration of anti-TNF-α reduced the $[Ca^{2+}]i$ level markedly, so did the pre-administration of 2-APB. We speculate that 2-APB acted on the renal vessel bed by inhibiting the IP_3Rs signaling pathway, and in turn increased GFR in rats with FHF while it has less effect on hepatic failure. These findings imply a crucial role of TNF-α in modulating IP_3R1 signaling activities and affecting GFR during FHF.

The findings strongly suggested that IP_3R1 over-expression in GMC and VSMC is an important molecular mechanism underlying TNF-α-induced GFR reduction.

There are many vasoconstrictor substances in patients with end-stage liver disease, and it is not plausible that inhibition of only one vasoconstrictive substance could efficiently block an increase in IP_3 and calcium levels within the GMC and VSMC.

Therefore, the terminal signaling pathway of the vasoconstrictor substances must be interrupted in order to achieve the blocking contractile effect of the vasoconstrictor substances on GMC and VSMC. Growing evidence indicates that TNF-α-dependent IP_3R1 up-regulation correlates with alterations in Ca^{2+} homeostasis in many conditions, and may be a specific target in future studies of renal dysfunction during end-stage liver disease.

In accordance with the data we have found, we propose that TNF-α plays a role in the development of renal dysfunction. Therefore, we hypothesize that it is possible that increased IP_3R1, induced by TNF-α, would lead to increased IP_3 sensitivity and Ca^{2+}-releasing activity in response to a variety of vasoconstrictors when HRS occurred.

Ryanodine Receptor in Kidney

There are two major Ca^{2+} release channels on the ER, the IP3R and the ryanodine receptor (RyR) [57, 58]. Three RyR isoforms have been described: RyR1 predominates in skeletal muscle, RyR2 in the heart [59], and RyR3 is expressed at low levels in various tissues [60]. With regard to the kidney, Tunwell and Lai [61] demonstrated that the RyR2 isoform is present in the rabbit kidney cortex. Most recently, RyR has been found in the afferent preglomerular arterioles [31, 42], proximal tubular cells [62], and human embryonic kidney cells [62].

Our research demonstrated that the RyR calcium channel did not play an important role in this process. A previous study found that blocking the release of calcium from the IP_3R can definitely protect the tubular cells against apoptosis, just as those in other organs [63, 64], but blocking the RyR cannot.

In conclusion, the findings in the present study demonstrate that TNF-α-induced reduction in GFR is responsible for the development of HRS. The enhanced constrictive activity of the renal vascular bed caused by vasoconstrictors due to overexpression of TNF-α-dependent IP_3R1 gene and an increase in glomerular calcium levels appears to contribute to GFR reduction.

The identification of TNFR1/PC-PLC/PKCα-dependent and TNFR2-dependent IP_3R1 genes may also help to elucidate the biological functions of these signal molecules more fully. The TNF-α-dependent IP_3R1 up-regulation may well be an important target in the treatment of HRS. The IP_3R might be a new target for the treatment of renal injury, especially for HRS patients.

References

[1] Berridge, M. J., Irvine, R. F. Inositol trisphosphate, a novel second messenger in cellular signal transduction. *Nature* 1984;312:315-321.

[2] Berridge, M. J., Irvine, R. F. Inositol phosphates and cell signalling. *Nature* 1989;341:197-205.

[3] Nishizuka, Y. The role of protein kinase C in cell surface signal transduction and tumour promotion. *Nature* 1984;308:693-698.

[4] Nishizuka, Y. The molecular heterogeneity of protein kinase C and its implications for cellular regulation. *Nature* 1988;334:661-665.

[5] Ferris, C. D., Huganir, R. L., Supattapone, S., Snyder, S. H. Purified inositol 1,4,5-trisphosphate receptor mediates calcium flux in reconstituted lipid vesicles. *Nature* 1989;342:87-89.

[6] Miyawaki, A., Furuichi, T., Maeda, N., Mikoshiba, K. Expressed cerebellar-type inositol 1,4,5-trisphosphate receptor, P400, has calcium release activity in a fibroblast L cell line. *Neuron* 1990;5:11-18.

[7] Berridge, M. J. Inositol trisphosphate and calcium signalling. *Nature* 1993;361:315-325.

[8] Furuichi, T., Yoshikawa, S., Miyawaki, A., Wada, K., Maeda, N., Mikoshiba, K. Primary structure and functional expression of the inositol 1,4,5-trisphosphate-binding protein P400. *Nature* 1989;342:32-38.

[9] Mignery, G. A., Newton, C. L., Archer, B. T., 3rd, Sudhof, T. C. Structure and expression of the rat inositol 1,4,5-trisphosphate receptor. *J. Biol. Chem.* 1990;265:12679-12685.

[10] Ross, C. A., Meldolesi, J., Milner, T. A., Satoh, T., Supattapone, S., Snyder, S. H. Inositol 1,4,5-trisphosphate receptor localized to endoplasmic reticulum in cerebellar Purkinje neurons. *Nature* 1989;339: 468-470.

[11] Otsu, H., Yamamoto, A., Maeda, N., Mikoshiba, K., Tashiro, Y. Immunogold localization of inositol 1, 4, 5-trisphosphate (InsP3) receptor in mouse cerebellar Purkinje cells using three monoclonal antibodies. *Cell Struct. Funct.* 1990;15:163-173.

[12] Nakagawa, T., Okano, H., Furuichi, T., Aruga, J., Mikoshiba, K. The subtypes of the mouse inositol 1,4,5-trisphosphate receptor are expressed in a tissue-specific and developmentally specific manner. *Proc. Natl. Acad. Sci. US* 1991;88:6244-6248.

[13] Hayashi, M., Monkawa, T., Saruta, T. Cell-specific expression of the IP (3) receptor gene family in the kidney. *Exp. Nephrol.* 2000;8:215-218.

[14] Monkawa, T., Hayashi, M., Miyawaki, A., Sugiyama, T., Yamamoto-Hino, M., Hasegawa, M., Furuichi, T., et al. Localization of inositol 1,4,5-trisphosphate receptors in the rat kidney. *Kidney Int.* 1998;53:296-301.

[15] Yang, T., Terada, Y., Nonoguchi, H., Tomita, K., Marumo, F. Expression of mRNAs for different types of IP3 receptors in rat kidneys. *Am. J. Physiol.* 1995;268:F1046-1052.

[16] Matsumoto, M., Nakagawa, T., Inoue, T., Nagata, E., Tanaka, K., Takano, H., Minowa, O., et al. Ataxia and epileptic seizures in mice

lacking type 1 inositol 1,4,5-trisphosphate receptor. *Nature* 1996;379: 168-171.

[17] Wu, D., Chen, X., Ding, R., Qiao, X., Shi, S., Xie, Y., Hong, Q., et al. Ischemia/reperfusion induce renal tubule apoptosis by inositol 1,4,5-trisphosphate receptor and L-type Ca2+ channel opening. *Am. J. Nephrol.* 2008;28:487-499.

[18] Rao, R. V., Ellerby, H. M., Bredesen, D. E. Coupling endoplasmic reticulum stress to the cell death program. *Cell. Death Differ.* 2004;11: 372-380.

[19] Rajekar, H., Chawla, Y. Terlipressin in hepatorenal syndrome: Evidence for present indications. *J. Gastroenterol. Hepatol.* 2011;26 Suppl. 1:109-114.

[20] Khattak, S., G KM, McMartin, K., Barrera, M., Kennedy, D., Koren, G. Pregnancy outcome following gestational exposure to organic solvents: a prospective controlled study. *JAMA* 1999;281:1106-1109.

[21] Gines, P., Guevara, M., Arroyo, V., Rodes, J. Hepatorenal syndrome. *Lancet* 2003;362:1819-1827.

[22] Arroyo, V., Fernandez, J., Gines, P. Pathogenesis and treatment of hepatorenal syndrome. *Semin. Liver Dis.* 2008;28:81-95.

[23] Wen, Y., Cui, W., Liu, P. Type I inositol 1, 4, 5-triphosphate receptors increase in kidney of mice with fulminant hepatic failure. *World J. Gastroenterol.* 2007;13:2344-2348.

[24] Demirtas, S., Can, M., Yarpuzlu, A. Hepatorenal syndrome. *Clin. Chem. Lab. Med.* 2006;44:379-386.

[25] Gerbes, A. L., Gulberg, V. [Renal impairment in liver diseases]. *Praxis* (Bern 1994) 2006;95:1535-1538.

[26] Qi, Z., Whitt, I., Mehta, A., Jin, J., Zhao, M., Harris, R. C., Fogo, A. B., et al. Serial determination of glomerular filtration rate in conscious mice using FITC-inulin clearance. *Am. J. Physiol. Renal. Physiol.* 2004;286: F590-596.

[27] Zhu, Z., Arendshorst, W. J. Angiotensin II-receptor stimulation of cytosolic calcium concentration in cultured renal resistance arterioles. *Am. J. Physiol.* 1996;271:F1239-1247.

[28] Wang, Y. R., Li, Z. G., Fu, J. L., Wang, Z. H., Wen, Y., Liu, P. TNFalpha-induced IP3R1 expression through TNFR1/PC-PLC/PKCalpha and TNFR2 signalling pathways in human mesangial cell. *Nephrol. Dial. Transplant.* 2011;26:75-83.

[29] Kuo, H. T., Shin, S. J., Kuo, M. C., Chen, H. C. Effects of specific endothelin-1 receptor antagonists on proliferation and fibronectin

production of glomerular mesangial cells stimulated with Angiotensin II. *Kaohsiung J. Med. Sci.* 2006;22:371-376.

[30] Hayashi, M., Monkawa, T., Yoshida, T., Sasamura, H., Matsumoto, M., Inoue, T., Mikoshiba, K., et al. Intracellular calcium concentration in the inositol trisphosphate receptor type 1 knockout mouse. *J. Am. Soc. Nephrol.* 1999;10:2094-2101.

[31] Fellner, S. K., Arendshorst, W. J. Angiotensin II Ca2+ signaling in rat afferent arterioles: stimulation of cyclic ADP ribose and IP3 pathways. *Am. J. Physiol. Renal. Physiol.* 2005;288:F785-791.

[32] Noll, G., Wenzel, R. R., Luscher, T. F. Endothelin and endothelin antagonists: potential role in cardiovascular and renal disease. *Mol. Cell. Biochem.* 1996;157:259-267.

[33] Pinto-Sietsma, S. J., Paul, M. A role for endothelin in the pathogenesis of hypertension: fact or fiction? *Kidney Int. Suppl.* 1998;67:S115-121.

[34] Miyauchi, T., Masaki, T. Pathophysiology of endothelin in the cardiovascular system. *Annu. Rev. Physiol.* 1999;61:391-415.

[35] Edwards, R. M., Trizna, W. Characterization of 125I-endothelin-1 binding to rat and rabbit renal microvasculature. *J. Pharmacol. Exp. Ther.* 1995;274:1084-1089.

[36] De Leon, H., Garcia, R. Characterization of endothelin receptor subtypes in isolated rat renal preglomerular microvessels. *Regul. Pept.* 1995;60:1-8.

[37] Pollock, D. M., Banks, R. O. Perspectives on renal blood flow autoregulation. *Proc. Soc. Exp. Biol. Med.* 1991;198:800-805.

[38] Gordienko, D. V., Clausen, C., Goligorsky, M. S. Ionic currents and endothelin signaling in smooth muscle cells from rat renal resistance arteries. *Am. J. Physiol.* 1994;266:F325-341.

[39] Hocher, B., Rohmeiss, P., Zart, R., Diekmann, F., Vogt, V., Metz, D., Fakhury, M., et al. Function and expression of endothelin receptor subtypes in the kidneys of spontaneously hypertensive rats. *Cardiovasc. Res.* 1996;31:499-510.

[40] Cao, L. Q., Banks, R. O. Cardiovascular and renal actions of endothelin: effects of calcium-channel blockers. *Am. J. Physiol.* 1990;258:F254-258.

[41] Pollock, D. M., Keith, T. L., Highsmith, R. F. Endothelin receptors and calcium signaling. *FASEB J.* 1995;9:1196-1204.

[42] Takenaka, T., Epstein, M., Forster, H., Landry, D. W., Iijima, K., Goligorsky, M. S. Attenuation of endothelin effects by a chloride channel inhibitor, indanyloxyacetic acid. *Am. J. Physiol.* 1992;262:F799-806.

[43] Schroeder, A. C., Imig, J. D., LeBlanc, E. A., Pham, B. T., Pollock, D. M., Inscho, E. W. Endothelin-mediated calcium signaling in preglomerular smooth muscle cells. *Hypertension* 2000;35:280-286.

[44] Yasuoka, H., Yoshizumi, M., Inui, D., Okishima, N., Houchi, H., Kirima, K., Oshita, S., et al. Effect of endothelin-1(1-31) on intracellular free calcium in cultured human mesangial cells. *Life Sci.* 1999; 65: PL267-272.

[45] Carswell, E. A., Old, L. J., Kassel, R. L., Green, S., Fiore, N., Williamson, B. An endotoxin-induced serum factor that causes necrosis of tumors. *Proc. Natl. Acad. Sci. US* 1975;72:3666-3670.

[46] Kriegler, M., Perez, C., DeFay, K., Albert, I., Lu, S. D. A novel form of TNF/cachectin is a cell surface cytotoxic transmembrane protein: ramifications for the complex physiology of TNF. *Cell* 1988;53:45-53.

[47] Kinouchi, K., Brown, G., Pasternak, G., Donner, D. B. Identification and characterization of receptors for tumor necrosis factor-alpha in the brain. *Biochem. Biophys. Res. Commun.* 1991;181:1532-1538.

[48] Chen, G., Goeddel, D. V. TNF-R1 signaling: a beautiful pathway. *Science* 2002;296:1634-1635.

[49] Schneider-Brachert, W., Tchikov, V., Neumeyer, J., Jakob, M., Winoto-Morbach, S., Held-Feindt, J., Heinrich, M., et al. Compartmentalization of TNF receptor 1 signaling: internalized TNF receptosomes as death signaling vesicles. *Immunity* 2004;21:415-428.

[50] Cunningham, A. J., Murray, C. A., O'Neill, L. A., Lynch, M. A., O'Connor, J. J. Interleukin-1 beta (IL-1 beta) and tumour necrosis factor (TNF) inhibit long-term potentiation in the rat dentate gyrus in vitro. *Neurosci. Lett.* 1996;203:17-20.

[51] Butler, M. P., O'Connor, J. J., Moynagh, P. N. Dissection of tumor-necrosis factor-alpha inhibition of long-term potentiation (LTP) reveals a p38 mitogen-activated protein kinase-dependent mechanism which maps to early-but not late-phase LTP. *Neuroscience* 2004;124:319-326.

[52] Stutzmann, G. E., Smith, I., Caccamo, A., Oddo, S., Laferla, F. M., Parker, I. Enhanced ryanodine receptor recruitment contributes to Ca2+ disruptions in young, adult, and aged Alzheimer's disease mice. *J. Neurosci.* 2006;26:5180-5189.

[53] Park, K. M., Yule, D. I., Bowers, W. J. Tumor necrosis factor-alpha-mediated regulation of the inositol 1,4,5-trisphosphate receptor promoter. *J. Biol. Chem.* 2009;284:27557-27566.

[54] Wang, J., Sun, J., Lu, S., Liu, P. [Tumor necrosis factor-alpha enhances type I inositol 1, 4, 5-triphosphate receptor expression in rat glomerular

afferent arterioles smooth muscle cells]. *Zhonghua Nei Ke Za Zhi* 2002; 41:86-89.

[55] Park, K. M., Yule, D. I., Bowers, W. J. Tumor necrosis factor-alpha potentiates intraneuronal Ca2+ signaling via regulation of the inositol 1,4,5-trisphosphate receptor. *J. Biol. Chem.* 2008;283:33069-33079.

[56] Lu, J. W., Wang, H., Yan-Li, J., Zhang, C., Ning, H., Li, X. Y., Zhang, H., et al. Differential effects of pyrrolidine dithiocarbamate on TNF-alpha-mediated liver injury in two different models of fulminant hepatitis. *J. Hepatol.* 2008;48:442-452.

[57] Mattson, M. P., Chan, S. L. Calcium orchestrates apoptosis. *Nat. Cell. Biol.* 2003;5:1041-1043.

[58] Marks, A. R. Intracellular calcium-release channels: regulators of cell life and death. *Am. J. Physiol.* 1997;272:H597-605.

[59] Fill, M., Copello, J. A. Ryanodine receptor calcium release channels. *Physiol. Rev.* 2002;82:893-922.

[60] Takeshima, H., Ikemoto, T., Nishi, M., Nishiyama, N., Shimuta, M., Sugitani, Y., Kuno, J., et al. Generation and characterization of mutant mice lacking ryanodine receptor type 3. *J. Biol. Chem.* 1996;271:19649-19652.

[61] Tunwell, R. E., Lai, F. A. Ryanodine receptor expression in the kidney and a non-excitable kidney epithelial cell. *J. Biol. Chem.* 1996;271: 29583-29588.

[62] Querfurth, H. W., Haughey, N. J., Greenway, S. C., Yacono, P. W., Golan, D. E., Geiger, J. D. Expression of ryanodine receptors in human embryonic kidney (HEK293) cells. *Biochem. J.* 1998;334 (Pt 1):79-86.

[63] Madesh, M., Hawkins, B. J., Milovanova, T., Bhanumathy, C. D., Joseph, S. K., Ramachandrarao, S. P., Sharma, K., et al. Selective role for superoxide in InsP3 receptor-mediated mitochondrial dysfunction and endothelial apoptosis. *J. Cell. Biol.* 2005;170:1079-1090.

[64] Groenendyk, J., Lynch, J., Michalak, M. Calreticulin, Ca2+, and calcineurin - signaling from the endoplasmic reticulum. *Mol. Cells* 2004; 17:383-389.

In: Inositol
Editors: H. Rocha and M. Cardoso
ISBN: 978-1-62808-520-4

Chapter 6

Pathophysiologic Significance of Inositol Phosphates (IPs) in Alzheimer's Disease

***Krishnaswamy Kannan*[1], *Rakhee Agarwal*[2] *and Nawab Ali*[3,*]**

[1]Department of Internal Medicine, Division of Rheumatology,
[2]Department of Pharmacology and Toxicology,
University of Arkansas for Medical Sciences, Little Rock, AR and
[3]Department of Applied Science, University of Arkansas at Little Rock,
Little Rock, AR, U.S.

Summary

Alzheimer's disease (AD) is the leading cause of adult onset dementia and the incidence is expected to rise in US, Japan and other developed countries. AD is characterized by the accumulation of amyloid–beta peptides forming plaques. Though the neuropathology of AD was discovered over a century ago, the molecular aspects of the disease development are unclear. It is speculated that amyloid-beta peptides target cholinergic neurons leading to

[*] Corresponding author: Nawab Ali, PhD, Associate Professor, Department of Applied Science, University of Arkansas at Little Rock, 2801 S University Avenue, Little Rock, Arkansas 72204. Phone: 501-569 8003, Fax: 501-569 8020, E-mail: nali@ualr.edu.

neural degeneration. Defective calcium signaling in neurons has been shown to play a critical role in the pathogenesis of AD. There is now increased interest and research on inositol phosphates (IPs) and calcium signaling since IPs are known to regulate Ca^{2+} homeostasis in endoplasmic reticulum, cell signaling, and cell death mechanisms. Thus, dysregulation of the IP mediated calcium homeostasis and associated cell physiology seems to play a pathological role in the development of many chronic illnesses particularly neurodegenerative diseases. In addition to IPs, aluminum toxicity also has been shown to contribute significantly to the perturbance of calcium metabolism, mitochondrial dysfunction, and inflammasome activation. There appears to be a convergence of dysregulated calcium metabolism, aluminum toxicity, and inflammasome activation leading to beta amyloid deposits and associated pathology in Alzheimer's. Similar to Alzheimer's, IPs also have been shown to influence the pathogenesis of other neurological diseases including bipolar disorder. At the cellular level, there might be common shared pathways and underlying mechanisms related to IP. In this review, the pathophysiologic significance of IPs in the development of AD and bipolar disorders is discussed.

History of Dementia

The history of dementia is as old as mankind. Historically, ancient Egyptians described the concept of dementia as a major memory disorder that occurs during aging. Plato and Horatius also thought that old age per se synonymous with senile dementia. Galen (130 - 201 AD) and Hippocrates, two physicians of classical antiquity with longstanding influence on Western Medicine thought that dementia was due to cerebral impairment which could be primary or secondary to other diseases in other organ systems. Alzheimer's disease (AD) was co discovered by two scientists, Aloe Alzheimer, a German psychiatrist and neuropathologist who first observed the characteristic neuropathology and described the first case of AD in 1906 and Emil Kraepelin, the head of the Psychiatry department at the University of Munich, who first identified the symptoms of the disease. The disease was named after the discoverer who presented the case at the 37th Meeting of the Southwest German Psychiatrists in Tübingen in 1906 (reviewed in [1]). Initially, the term AD was used to distinguish a pre-senile dementia from the general paresis, a paralytic dementia. Later on, the possibility that pre-senile Alzheimer's was clinically or histologically indistinguishable from senile dementia was clarified in late 1970's.

Till the year 1975, AD was thought to be a rare condition and only in the year 1977 the first conference on Alzheimer's was held and subsequently the clinical interest in the disease expanded into a subject for academic medicine. The identification and clinical characterization of Alzheimer's disease at the onset of the 20th century was a turning point for the understanding of senile dementia. The concepts and histological findings presented by the early researchers of Alzheimer's disease remain relevant today.

Currently there is no cure for AD, though non-specific palliative and other therapies are being attempted. The onset of AD and its course is often variable, thus left undiagnosed for months to years [1]. As the disease advances, common symptoms can include decline in cognitive intelligence, confusion, irritability, aggression, mood swings, trouble with language, and long-term memory loss. Thus, cognitive decline and AD pose a significant burden not only on affected individuals, but also on their caregivers and society at large.

Approximately 100 billion US dollars are spent towards the healthcare of AD patients every year towards long-term home and hospital care [2, 3]. As per the recent estimate by Alzheimer's Foundation of America (2012), 5.1 million Americans suffered from this disease. It is anticipated that advances in molecular medicine and new research data will soon lead to early diagnosis and better therapies.

Introduction

AD is a progressive neurodegenerative disorder which is clinically characterized by a widespread functional disturbance in the human brain because of loss of neurons, deposition of beta-amyloid peptide (Aβ) in cerebral blood vessels, development of plaques and the presence of neurofibrillary tangles. Loss of specific neural circuits in the neocortex, hippocampus, and basal forebrain cholinergic system is thought to be an additional reason for in cognitive decline in AD patients [4]. Further, the degeneration of synapses and neuronal death results in reduction in the temporal and frontal lobes in brain, which associated with learning and memory processes [5]. During early stages of AD, the disease progresses with neuropsychiatric symptoms including gradual memory loss, decline in normal activity, and behavioral changes, while at later stages patients suffer more pervasive loss of recent memory, impairment in coordinated body movements, well known skills, language, decision making, etc. The presence of amyloid plaques in the brain at the time of autopsy is the definitive diagnosis of AD.

Calcium ions (Ca^{2+}) have been shown to play a pivotal role in the maintenance of nerve cell circuits in the brain. The cellular organelle endoplasmic reticulum (ER) not only serves as a reservoir for Ca^{2+}, but also helps to maintain Ca^{2+} homeostasis, in which mitochondria play an active role [5-8]. Involved in this signal cascade are ER membrane receptor channels and inositol triphosphate (IP_3), the lipid derived signaling molecule, which is also considered 'secondary messenger' in the cell. Under conditions of calcium overload in the ER, activation of biochemical cascades leads to impaired mitochondrial respiration, altered energy output, and oxidative stress and in certain conditions activation of inflammasome and other cellular changes occur [6].

Besides being an age-associated disease, AD seems to be influenced by genetic factors as well. Several experimental models of AD have provided evidence in support of a causal link between perturbed ER Ca^2 homeostasis and cognitive deficits and neurodegeneration in AD [5, 7]. Sherrington et al. [8] have shown a strong association between mutations of presenilin-1 (*PS1*), an early-onset familial AD (FAD) gene which is implicated in dysregulation of Ca^2 homeostasis.

There is now mounting evidence for additional pathophysiological factors including accumulation of Aβ, increased oxidative stress, mitochondrial dysfunction, and aberrant inflammasome activation. Collectively, all these changes seem to accelerate both aging and dementia in patients diagnosed with AD. However, there are major gaps in knowledge in our understanding of the pathogenesis of AD, particularly the formation and accumulation of Aβ deposits in neuronal cells and the role of IP_3 metabolism and toxicity during this process. Thus, a better understanding of the underlying biochemical mechanisms of IP_3 dysregulation leading to perturbance of calcium metabolism could open up new avenues for developing novel therapies for this disease. Similarly, an enlarged view of the underlying mechanisms of IP_3 metabolism, aluminum toxicity, and its impact in the development of AD would help to identify novel drug targets and develop therapeutic strategies to halt the progress of the disease. In this review, the current scientific advances on the causes, risk factors, diagnostic tools, and the role of inositol and inositol phosphates and a list of potential therapeutic targets were critically analyzed and discussed.

Causes of the Disease

AD is a neurodegenerative disorder characterized by the presence of extracellular senile plaques and intracellular neurofibrillary tangles in the brains of the affected individuals. These plaques are composed primarily of short peptides known as amyloid β-peptides (Aβ), which can be found in a diffused form or in a more organized form called senile or neuritic plaques. The major protein subunit A4 of Aβ peptides is an insoluble, highly aggregating small polypeptide of relative molecular mass 4,500 Da [6, 9, 10]. Aβ are derived from the proteolytic processing of the β-amyloid precursor protein (APP) [11], which is a normal constituent of neural and non-neural cells [12]. Deposits of Aβ complexes are found outside as well as inside nerve cells and in the walls of small blood vessels in the brain [13]. Neurofibrillary tangles are formed by hyperphosphorylation of a microtubule-associated protein known as Tau, causing it to aggregate, or group, in an insoluble form. Although pre-mortem diagnoses of this disease is not available, these Tau-containing neurofibrillary tangles are used to track severity of the disease. A brain autopsy is needed for the definitive diagnosis of Alzheimer's disease.

In recent years, the molecules and biochemical pathways that mediate the inflammation associated with Alzheimer's disease have received considerable attention [14-16]. Importantly, microglial activation in response to injury, illness, aging or other unknown causes triggers a cascade of biochemical events with overexpression of interleukin-1 (IL-1) and other proinflammatory cytokines, which has been increasingly implicated in the pathogenesis of AD [14-17]. These results suggest a pivotal role for inflammasome activation and associated pathologies. Recently, Vom Berg et al. demonstrated an active role for IL-1 using a mouse model of AD [18]. Oxidative stress, dysregulated calcium and energy metabolism followed by beta amyloid deposits and inflammasome activation contribute to neurodegenerative process and associated cognitive deficits (5,14). Dysregulation of IP metabolism and accumulation of aluminum in neuronal cells also contribute to the pathogenesis of the disease process (5,17,18). Growing evidence also indicates that mitochondrial damage and dysfunction are critical factors in age-related neurodegenerative diseases including AD (20,24-26). At present, there is an imperfect understanding of the mechanisms involved, though the role of neuroinflammation in AD pathogenesis is gaining ground.

Hypotheses on the Causes of Alzheimer's

Neurofibrillary tangles (NFTs) and amyloid plaques are the two defining pathological hallmarks of AD. Several hypotheses have been put forward to explain these molecular events and subsequent neuropsychiatric symptoms. The oldest of these is the cholinergic hypothesis, which suggests that the main causative factor in AD is deficiency of the key neurotransmitter acetylcholine. Most first generation therapies based on this hypothesis could alleviate a few symptoms of the disease, but could not halt or cure Alzheimer's [16, 19]. Two other hypotheses suggest that either tau protein or amyloid beta initiates the cascade. The predominant model of pathogenesis in AD is the amyloid cascade hypothesis, but this does not fully account for the molecular events observed. Thus, the "tau and tangle" hypothesis was put forth to explain the pathology of neurofibrillary tangles [19]. Accumulating evidence in recent years has shown that amyloid-beta peptide drives cerebral neuroinflammation by activating microglia. Furthermore, amyloid-beta activation of NLRP3 inflammasome in microglia is now believed to be fundamental for IL-1β maturation and subsequent inflammatory events [20].

Genetics

Despite decades of research, the genetic basis of AD is still evolving and the molecular genetic analyses suggest that multiple genes may influence one's susceptibility to AD. The vast majority of cases of AD are sporadic – they do not run in families [21]. Nevertheless, genes for apolipoprotein E4 isoform are at increased risk of AD, though the underlying mechanisms are not clear [22]. Genetic analysis of Late-Onset Alzheimer Disease (LOAD), the most common cause of dementia has shown a 20% of the disease-attributable risk to ϵ4 variant in APOE [23]. Further case control studies are necessary to arrive at better conclusions.

Risk Factors of Alzheimer's

Age is one of the dominant risk factors of AD. Alzheimer's is now regarded as the fourth most common cause of deaths in the elderly, affecting nearly 2% of the population in industrialized countries. The risk of

Alzheimer's dramatically increases among senior citizens over 70 years of age. Incidence of Alzheimer's is higher among women since they have a longer life span [21, 25, 28]. Accumulation of Aβ during normal aging may accelerate with increased oxidative stress, cellular energy deficits due to mitochondrial dysfunction and inflammation have been reported to increase the risk of developing AD; these and other risk factors are summarized in Table-1 [21, 24-28]. Other causes are acetylcholine depletion, free radical damage, lipofuscin, inflammation, biopterin, homocysteine, advanced glycation end products and aluminum toxicity. Clinical tests for folic acid, vitamin E, estrogens, statins, and NSAIDs may reduce the number of Alzheimer's patients by half by delaying the onset of the disease by 5 yrs.

Table 1. Risk Factors in Alzheimer's disease

Major Risk Factors:

- Advanced Age 65 to 90 years
- PS1, PS2, APP, A2M mutations
- ApoE4

Additional Risk Factors:

- Sedentary life style
- Diabetes and associated metabolic disorders
- Traumatic brain injury or trauma
- Chronic oxidative stress
- Clinical depression

Reference: [7, 26, 27]

Computed Tomography and Magnetic Resonance in the Diagnosis

The National Institute on Aging (NIA) criteria of Alzheimer's is based on the patient age and clinical history. Besides the clinical diagnosis of neurological basis of cognitive decline among patients with AD, structural neuroimaging such as computed tomography (CT) or magnetic resonance (MR) are also routinely employed for the diagnosis of memory disorders [21, 29]. Additionally, functional imaging modalities such as positron emission tomography (PET), single photon emission CT (SPE-CT), and functional MR imaging which show physiologic changes in brain are better options available to identify any subtle pathologic change in brain at early stages of Alzheimer's.

Pathophysiological Roles of Inositol Phosphates in Neurologic Diseases

Inositol phosphates are well known signaling molecules which are involved in many biological processes including calcium signaling, energy metabolism, cell growth, and cell death mechanisms. In particular, inositol-1,4,5-triphosphate (IP_3) has been best characterized as a second messenger releasing intracellular calcium. In recent years, inositol pyrophosphates (PP-IPs) have emerged as important signaling molecules in eukaryotic cells [6, 9]. They also serve as second messengers, stress response molecules, and as critical regulators of cell growth [10].

Based on extensive experimental evidence, the physiological functions of IPs have been placed under the following categories: (1) Nuclear export of RNA, (2) vesicular trafficking, endocytosis and cell morphology, (3) mitochondrial integrity and function, energy output, (4) cell cycle and cell death mechanisms, and (5) growth factor and cytokine signaling. Studies, primarily from the laboratories of Daniel Lindner and Solomon Snyder have significantly expanded our current understanding of the role of inositol pyrophosphates in cell death and its regulation [9, 14, 15].

Most inositol in nature exists as membrane bound phosphatidylinositol (PI) phospholipid which is phosphorylated to phosphatidylinositol-4,5-bisphosphate (PIP_2). Upon receptor activation, heterotrimeric G-protein subunit Gq is activated which couples receptor signaling response to the activation of an effector enzyme, PI-specific phospholipase C (PI-PLC). Activated PI-PLC hydrolyzes PIP2 into diacylglycerol (DAG) and cytosolic inositol(1,4,5)P_3 or IP_3. DAG stays in the membrane and activates protein kinase C whereas IP_3 binds to a receptor on endoplasmic reticulum (ER) and releases calcium ions from ER storage. Calcium also participates with DAG to stimulate the activity of protein kinase C for protein phosphorylation. Heterotrimeric G-proteins serve as intermediate that link cell surface receptors to PI-PLC [30, 31]. Inositol phosphates thus regulate cellular calcium. The second messenger function of IP_3 is terminated by its rapid hydrolysis to inositol bis-phosphate and then to inositol monophosphate and ultimately to inositol by inositol monophosphate phosphatase (IMPase), a reaction that is inhibited by lithium therapy. Inositol is recycled back to membrane for phosphoinositol biosynthesis and IP_3 is further phosphorylated to inositol tera-kis (IP_4), inositol penta-kis (IP_5) and inositol hexa-kis phosphates (IP_6). As mentioned elsewhere in this review, more phosphorylated inositol phosphates (IP_7, IP_8 and possibly IP_9) than IP_6 have now been isolated and their physiological functions are being studied. The physiological functions of inositol phosphates are predicted based on their interactions with proteins or regulation of cellular processes. These functions vary from their role in vesicular trafficking, nuclear export of mRNA, chromatin remodeling, cell death and differentiation to glucose metabolism by altering sensitivity to insulin receptor.

Higher inositol phosphates (inositol phosphate pyrophosphates; IP_7 and IP_8) have been shown to bind to adapter proteins of clathrin coated vesicles and coatomer complexes that participate in vesicular membrane transport [32]. Importantly, these molecules also bind to neuronal synaptic adapter protein AP-180 [33] which plays a role in synaptic transmission, a process compromised in AD. The assembly of synaptic vesicle is dissociated due to the binding of these inositol phosphates. Such roles of inositol phosphate molecules clearly suggest their involvement in neuronal functions and therefore the likelihood that they play a role in neurological disorders including Alzheimer's disease. Since this monograph contains exclusive chapters on inositol and IP, the discussion in this chapter is mainly devoted to the pathophysiologic role of IP in neurologic diseases such as Alzheimer's.

Innate Immunity, Inflammasome Activation, and Autophagy Mechanisms

Under steady state conditions, innate immune responses including inflammasome activation provide protective function in various organ systems of the host which includes immune-privileged sites such as eye and brain [34-37]. However, excess activation of inflammasome activation can result in dysregulated immune response resulting in adverse reactions and in the case of Alzheimer's patients, a life threatening disorder [20, 38-41]. During NLRP3 inflammasome activation, precursors of pro-inflammatory cytokines IL-1β and IL-18 are cleaved to mature and biologically active forms of cytokines by caspase-1 enzyme. When secreted out by innate immune cells, bioactive IL-1β and IL-18 can amplify the local inflammatory process in the brain [40, 41]. In summary, amyloid oligomers and fibrils are sensed as endogenous danger signals by the host brain cells and activate NLRP3 inflammasomes, probably via K^+ efflux in neurons and via lysosomal damage in microglia [41]. When this inflammasome activation becomes a non-resolving inflammatory process, it can result in a chronic disease such as Alzheimer's [39-42].

To thrive under various stress conditions, cells use various intracellular and extracellular stimuli and biochemical adaptive mechanisms. Importantly, mechanistic target of rapamycin complex 1 (mTORC1) plays a critical role in regulating growth and metabolism by integrating signals from the cellular environment and promoting approximately 650 genes involved in ribosome and protein synthesis [43]. Acting in parallel with the TORC1 pathway, PPIPs control the activity of the class I histone deacetylase Rpd3L. Thus, the PP-IP second messengers simultaneously activate and tune the global response to stress and starvation signals [43].

Aluminum Toxicity, Inositol Phosphates and Susceptibility to Alzheimer's

Aluminum is the third most abundant metal on the earth and a non-essential element to the biological system. Aluminum is also a neurotoxin which is also causally linked to neurodegenerative diseases. Most human exposure to aluminum occurs via food chain and environmental interactions; high aluminum containing foods include tea, herbs, spices, grain products, processed cheeses and salt. The first direct evidence linking aluminum to Alzheimer's came from the study published in 1973 in which significant accumulation of aluminum was demonstrated in the brain tissues of Alzheimer's patients [44]. Aluminum was directly injected into the animal brain leading to the formation of neurofibrillary tangles suggesting a direct role for aluminum [45, 46]. Environmental contamination of aluminum in drinking water in Camelford, UK in 1988 provided further support to the 'aluminum hypothesis' for cognitive impairment. Additional studies in Europe suggested that aluminum has adverse effects on human memory function and cause dementia when it is accumulated in brain tissues (reviewed in [47]).

Aluminum has been shown to inhibit Ca^{2+} sequestration into mitochondria and microsomes. Additionally, deficiencies of acetylcholine and neuronal choline acetyltransferase in the neuronal tissue are considered contributing factors of neurotoxicity by aluminum [48]. The binding of aluminum with phosphate groups in inositol polyphosphates affects its function in second messenger system [49, 50].

In vitro studies have shown that amyloid ß-peptides enhance the binding of IP_3 and IP_4 to their receptors [51]. Studies by Galila showed that there is 10-35% elevation in free inositol levels in occipital and parietal cortex of Alzheimer's disease patient as compare to controls [52]. It is shown through many studies that high level of myo-inositol also promotes the formation of amyloid plaques and hence Alzheimer's disease. In hippocampal and cortical slices from rat brain, aluminum decreases muscarinic, adrenergic, and metabotropic receptor stimulated phosphoinositide hydrolysis [53].

At the cellular level, beta amyloid oligomers have been shown to activate inflammasome thus linking Aβ to chronic inflammation and AD pathology [20, 41]. Aβ oligomers and fibrils can trigger innate immune response via inflammasome activation resulting in non-resolving inflammatory and apoptotic responses in human brain [20, 35, 41, 45, 54]. Collectively, these studies demonstrate a close association between altered IP metabolism and aluminum toxicity which in an aging brain promoted increased susceptibility to AD. Further studies are needed to better understand the underlying mechanisms involved.

Lithium Therapy: Disruption in Inositol Accumulation

Similar to Alzheimer's, calcium metabolism has been shown to play a critical role in the pathogenesis of other neurological diseases including bipolar disorder [53-57]. Hexahydroxycyclohazane or inositol is one of the dietary nutrients exists in nine possible spacial isomers. Among these, the D-myo-inositol is most prevalent in nature followed by D-scyllo-inositol. In brain, the D-myo-inositol is present at relatively higher concentrations compared to other tissues. It is transported actively by Na+/myo-inositol transporter and serves as osmolyte. It is phosphorylated to myo-inositol-3-phosphate by D-glucose-6-synthase and dephosphorylated to myo-inositol by a inositol phosphate monophosphatase (IMPase). Recycling of inositol takes place by way of its incorporation into phosphatidylinositol phospholipid within the cell membrane where it is further phosphorylated in a step wise phosphorylation reaction giving rise to phosphatidylinositol (4,5)-bisphopsphate, a substrate for receptor activated phosphatidylinositol specific phospholipase C that results in the formation of water soluble inositol-(1,4,5)-trisphosphate (IP_3) second messenger for calcium mobilization [54-60].

Lithium inhibits IMPase and shifts metabolism towards synthesis of other inositol phosphates including accumulation of IP_3 which would otherwise be dephosphorylated by highly active inositol phosphatases terminating second messenger function of IP_3. This property of lithium is commonly used in laboratory experiments studying cellular accumulations of inositol phosphates and their metabolism. The therapeutic effects of lithium treatment in bipolar and other psychiatric disorders are hypothesized to be the result of IMPase inhibition [55-58].

Inositol and Inositol Phosphates as New Drugs for Alzheimer's Disease

Attempts have been made to develop psychiatric drugs based on actions of lithium that would inhibit IMPase yet cause less lithium related side effects. However, understanding of the exact mechanism of action of lithium and therefore lithium-based drug is limited. Nevertheless, the pharmaceutical industry, particularly nature based herbal medicine industry, is poised to market inositol as dietary supplement or drug for the treatment of Alzheimer's disease while the clinical trials for safety evaluation are still awaiting results.

Post-mortem studies have shown that the brain tissues of Alzheimer's patients have a deficiency of essential fatty acids in one of the principal classes of phospholipids. Aluminum build-up may be prevented by dietary administration [59] which is considered as one of the possible pathogenic factors into the brain [59]. A double blind, crossover placebo controlled trial with Alzheimer's patient using 6 grams of inositol daily for a month resulted in improved language and orientation with no serious side effects [59]. When higher doses are administered (12 gram) for a month in a double blind and crossover study compared to placebo, there was a significant benefit for Hamilton depression scale (HDS), frequency and severity of panic attacks and significant decline in severity of agoraphobia and reduction in scores of OCD symptoms. There were no changes in hematology, kidney or liver functioning reported suggesting minimal side effects. This study shows that inositol has a therapeutic effect towards depression, panic and OCD. However, any acceptable therapeutic effects of inositol in Alzheimer's treatment were not evident. Therefore, it is currently not clear whether inositol is useful for Alzheimer's treatment. Based on its biological significance such as its precursor role in phosphatidylinositol cycle, a further controlled clinical trial of inositol in an Alzheimer's patient is required to realize its potential therapeutic application and at least its precautionary use in general treatment of bipolar disorders. The therapeutic effects of Lithium treatment in bipolar disorder hypothesized to be result of IMPase inhibition (53-55).

Future Directions and Conclusion

A large body of evidence has shown that IP_3 and other IPs are important signaling molecules and will continue to be useful in the study of neurobiology and in our better understanding of neurological and neurodegenerative diseases. In addition, the importance of calcium signaling is becoming evident in neuroinflammation and neurodegenerative diseases. Calcium homeostasis in particular has been shown to be essential for normal activation of inflammasome and immune function. The literature reviewed here has revealed new roles for calcium signaling in brain neurons and in particular Alzheimer's. In stressed cells, the intracellular calcium machinery becomes sensitized, resulting in increased calcium release from the ER into the cytosol. This calcium signal can lead to stimulation of protective biochemical defenses including autophagy. However, under pathological stress conditions, aberrant calcium signaling occurs that can trigger deleterious physiological outcomes. The relation between ER stress and calcium homeostasis appears complex. A better understanding of inositol phosphate signaling in ER stress, mitochondrial function, metabolomics and neuroinflammation may allow the discovery of new and better targets which may pave the way for novel therapies for patients diagnosed with Alzheimer's and other disorders. In this direction, more studies are required to further our understanding of pathophysiological roles of inositol phosphates especially of higher inositol phosphates including IP_6, IP_7 and IP_8. Therefore, a recent attempt to explore therapeutic potential of the most abundant inositol hexakisphosphate should be considered a step in right direction to expand future research to develop new approaches to treat Alzheimer's disease [61].

References

[1] Boller, F. & Forbes, M. M. (1998). History of dementia and dementia in history, an overview. *J Neurol Sci. 158*, 125-133.

[2] Bloom, B. S., de, P. N. & Straus, W. L. (2003). Cost of illness of Alzheimer's disease, how useful are current estimates? *Gerontologist. 43*, 158-164.

[3] Small, G. W., Rabins, P. V., Barry, P. P., Buckholtz, N. S., DeKosky, S. T., Ferris, S. H., Finkel, S. I., Gwyther, L. P., Khachaturian, Z. S., Lebowitz, B. D., McRae, T. D., Morris, J. C., Oakley, F., Schneider, L. S., Streim, J. E., Sunderland, T., Teri, L. A. & Tune, L. E. (1997). Diagnosis and treatment of Alzheimer disease and related disorders. Consensus statement of the American Association for Geriatric Psychiatry, the Alzheimer's Association, and the American Geriatrics Society. *JAMA. 278*, 1363-1371.

[4] Price, D. L., Sisodia, S. S. & Borchelt, D. R. (1998). Alzheimer disease--when and why? *Nat Genet. 19*, 314-316.

[5] Mattson, M. P. (2007). Calcium and neurodegeneration. *Aging Cell. 6*, 337-350.

[6] Masters, C. L., Simms, G., Weinman, N. A., Multhaup, G., McDonald, B. L. & Beyreuther, K. (1985). Amyloid plaque core protein in Alzheimer disease and Down syndrome. *Proc Natl Acad Sci U S A. 82*, 4245-4249.

[7] Price, D. L., Sisodia, S. S. & Borchelt, D. R. (1998). Genetic neurodegenerative diseases, the human illness and transgenic models. *Science. 282*, 1079-1083.

[8] Sherrington, R., Froelich, S., Sorbi, S., Campion, D., Chi, H., Rogaeva, E. A., Levesque, G., Rogaev, E. I., Lin, C., Liang, Y., Ikeda, M., Mar, L., Brice, A., Agid, Y., Percy, M. E., Clerget-Darpoux, F., Piacentini, S., Marcon, G., Nacmias, B., Amaducci, L., Frebourg, T., Lannfelt, L., Rommens, J. M. & St George-Hyslop, P. H. (1996). Alzheimer's disease associated with mutations in presenilin 2 is rare and variably penetrant. *Hum Mol Genet. 5*, 985-988.

[9] Masters, C. L., Multhaup, G., Simms, G., Pottgiesser, J., Martins, R. N. & Beyreuther, K. (1985). Neuronal origin of a cerebral amyloid, neurofibrillary tangles of Alzheimer's disease contain the same protein as the amyloid of plaque cores and blood vessels. *EMBO J. 4*, 2757-2763.

[10] Amtul, Z., Uhrig, M., Wang, L., Rozmahel, R. F. & Beyreuther, K. (2012). Detrimental effects of arachidonic acid and its metabolites in cellular and mouse models of Alzheimer's disease, structural insight. *Neurobiol Aging. 33*, 831.e21-31.

[11] Esch, F. S., Keim, P. S., Beattie, E. C., Blacher, R. W., Culwell, A. R.,Oltersdorf, T., McClure, D. & Ward, P.J. (1990). Cleavage of amyloid beta peptide during constitutive processing of its precursor. *Science.* 248, 1122-1124.

[12] Haass, C., Schlossmacher, M.G., Hung, A.Y., Vigo-Pelfrey,C., Mellon, A., Ostaszewski, B.L., Lieberburg, I., Koo, E.H., Schenk, D., Teplow, D.B., et al. (1992). Amyloid beta-peptide is produced by cultured cells during normal metabolism. *Nature.* 359, 322-325.

[13] Katzman, R., Brown, T., Fuld, P., Peck, A., Schechter, R. & H. Schimmel. (1983). Validation of a short Orientation-Memory-Concentration Test of cognitive impairment. *Am J Psychiatry.* 140, 734-739.

[14] Griffin, W. S. (2013). Neuroinflammatory cytokine signaling and Alzheimer's disease. *N Engl J Med. 368*, 770-771.

[15] Mrak, R. E. & Griffin, W. S. (2005). Potential inflammatory biomarkers in Alzheimer's disease. *J Alzheimers. Dis. 8*, 369-375.

[16] Griffin, W. S., Liu, L., Li, Y., Mrak, R. E. & Barger, S. W. (2006). Interleukin-1 mediates Alzheimer and Lewy body pathologies. *J Neuroinflammation. 3*:5.

[17] Griffin, W. S. (2006). Inflammation and neurodegenerative diseases. *Am J Clin Nutr. 83*, 470S-474S.

[18] Vom, B. J., Prokop, S., Miller, K. R., Obst, J., Kalin, R. E., Lopategui-Cabezas, I., Wegner, A., Mair, F., Schipke, C. G., Peters, O., Winter, Y., Becher, B. & Heppner, F. L. (2012). Inhibition of IL-12/IL-23 signaling reduces Alzheimer's disease-like pathology and cognitive decline. *Nat Med. 18*, 1812-1819.

[19] Mudher, A., & Lovestone, S. (2002). Alzheimer's disease-do tauists and baptists finally shake hands? *Trends Neurosci.* 25, 22-26.

[20] Heneka, M.T., Kummer, M.P., Stutz, A., Delekate,A., Schwartz, S., Vieira-Saecker, A., Griep, A., Axt, D., Remus, A., Tzeng, T.C., Gelpi, E., Halle, A., Korte, M., Latz, E., & Golenbock, D.T. (2013). NLRP3 is activated in Alzheimer's disease and contributes to pathology in APP/PS1 mice. *Nature.* 493, 674-678.

[21] Mattson, M. P. (2004). Pathways towards and away from Alzheimer's disease. *Nature. 430*, 631-639.

[22] Roses, A. D. (1997). A model for susceptibility polymorphisms for complex diseases, apolipoprotein E and Alzheimer disease. *Neurogenetics. 1*, 3-11.

[23] Sadigh-Eteghad, S., Talebi, M. & Farhoudi, M. (2012). Association of apolipoprotein E epsilon 4 allele with sporadic late onset Alzheimer`s disease. A meta-analysis. *Neurosciences (Riyadh). 17*, 321-326.

[24] Mattson, M. P. (2002). Oxidative stress, perturbed calcium homeostasis, and immune dysfunction in Alzheimer's disease. *J Neurovirol. 8*, 539-550.

[25] Mattson, M. P. (2012). An ECSIT-centric view of Alzheimer's disease. *Bioessays. 34*, 526-527.

[26] Yang, J. L., Weissman, L., Bohr, V. A. & Mattson, M. P. (2008). Mitochondrial DNA damage and repair in neurodegenerative disorders. *DNA Repair (Amst). 7*, 1110-1120.

[27] Tang, J., Pei, Y. & Zhou, G. (2013). When Aging-Onset Diabetes Is Coming Across With Alzheimer Disease, Comparable Pathogenesis And Therapy. *Exp Gerontol. pii: S0531-5565(13)00120-4.*

[28] Rodrigue, K. M., Rieck, J. R., Kennedy, K. M., Devous, M. D., Diaz-Arrastia, R. & Park. D. C. (2013). Risk Factors for beta-Amyloid Deposition in Healthy Aging, Vascular and Genetic Effects. *JAMA Neurol. 70*:600-606.

[29] Lowe, V. J., Peller, P. J., Weigand, S. D., Montoya, Q. C., Tosakulwong, N., Vemuri, P., Senjem, M. L., Jordan, L., Jack, C. R., Jr., Knopman, D. & Petersen, R. C. (2013). Application of the National Institute on Aging-Alzheimer's Association AD criteria to ADNI. *Neurology. 80:2130-37.*

[30] Birnbaumer, L., Abramowitz, J., Yatani, A., Okabe, K., Mattera, R., Graf, R., Sanford, J., Codina, J. & Brown, A. M. (1990). Roles of G proteins in coupling of receptors to ionic channels and other effector systems. *Crit Rev Biochem Mol Biol. 25*, 225-244.

[31] Hall, A. (1992). Signal transduction through small GTPases--a tale of two GAPs. *Cell. 69*, 389-391.

[32] Ali, N., Duden, R., Bembenek, M. E. & Shears, S. B. (1995). The interaction of coatomer with inositol polyphosphates is conserved in Saccharomyces cerevisiae. *Biochem J. 310*, 279-284.

[33] Ye, W., Ali, N., Bembenek, M. E., Shears, S. B. & Lafer, E. M. (1995). Inhibition of clathrin assembly by high affinity binding of specific inositol polyphosphates to the synapse-specific clathrin assembly protein AP-3. *J Biol Chem. 270*, 1564-1568.

[34] Kleinman, M. E. & Ambati. J. (2012). A window to innate neuroimmunity, Toll-like receptor-mediated cell responses in the retina. *Adv Exp Med Biol. 723,*3-9.

[35] Nathan, C. & Ding. A. (2010). Nonresolving inflammation. *Cell. 140,* 871-882.

[36] Lamkanfi, M. (2011). Emerging inflammasome effector mechanisms. *Nat Rev Immunol. 11*, 213-220.

[37] Lamkanfi, M. & Dixit, V. M. (2012). Inflammasomes and their roles in health and disease. *Annu Rev Cell Dev Biol. 28*,137-61.

[38] Liu, S. B., Mi, W. L. & Wang, Y. Q. (2013). Research progress on the NLRP3 inflammasome and its role in the central nervous system. *Neurosci Bull. (Epub)*, PMID 23512739.

[39] Goldmann, T., Tay, T. L. & Prinz. M. (2013). Love and death, microglia, NLRP3 and the Alzheimer's brain. *Cell Res. 23*, 595-596.

[40] Salminen, A., Ojala, J., Suuronen, T., Kaarniranta, K. & Kauppinen. A. (2008). Amyloid-beta oligomers set fire to inflammasomes and induce Alzheimer's pathology. *J Cell Mol Med. 12*, 2255-2262.

[41] Halle, A., Hornung, V., Petzold, G. C., Stewart, C. R., Monks, B. G., Reinheckel, T., Fitzgerald, K. A., Latz, E., Moore, K. J. & Golenbock, D. T. (2008). The NALP3 inflammasome is involved in the innate immune response to amyloid-beta. *Nat Immunol. 9*, 857-865.

[42] Trinh, N. H., Hoblyn, J., Mohanty, S. & Yaffe, K. (2003). Efficacy of cholinesterase inhibitors in the treatment of neuropsychiatric symptoms and functional impairment in Alzheimer disease, a meta-analysis. *JAMA. 289*, 210-216.

[43] Worley, J., Luo, X. & Capaldi,A.P. (2013). Inositol Pyrophosphates Regulate Cell Growth and the Environmental Stress Response by Activating the HDAC Rpd3L. *Cell Rep.* 3,1476-1482

[44] Crapper McLachlan, D. R., Dalton, A. J., Kruck, T. P., Bell, M. Y., Smith, W. L., Kalow, W. & Andrews, D. F. (1991). Intramuscular desferrioxamine in patients with Alzheimer's disease. *Lancet. 337*, 1304-1308.

[45] Klatzo, I., Wisniewski, H. & Streicher. E. (1965). Experimental production of neurofibrillary degeneration. I. Light microscopic observations. *J Neuropathol. Exp Neurol. 24*,187-99.

[46] Terry, R. D. & Pena. C. (1965). Experimental production of neurofibrillary degeneration 2. Electron microscopy, phosphatase histochemistry and electron probe analysis. *J Neuropathol. Exp Neurol. 24*, 200-10.

[47] Kawahara, M. & Kato-negishi, M. (2011). Link between aluminum and the pathogenesis of Alzheimer's disease: The integration of the aluminum and amyloid cascade hypotheses. *Int. J. Alzheimers Dis.* 2011:276393.

[48] King, R. G., Sharp, J. A. & Boura, A. L. (1983). Aluminium, choline transportation and Alzheimer's disease. *Med J Aust. 2*, 606-607.

[49] Collins, T. J., Lipp, P., Berridge, M. J. & Bootman, M. D. (2001). Mitochondrial Ca(2+) uptake depends on the spatial and temporal profile of cytosolic Ca(2+) signals. *J Biol Chem. 276*, 26411-26420.

[50] Berridge, M. J. (1986). Inositol trisphosphate and calcium mobilization. *Ciba Found. Symp. 122*, 39-57.

[51] Cowburn, R. F., Wiehager, B. & Sundstrom. E. (1995). beta-Amyloid peptides enhance binding of the calcium mobilising second messengers, inositol(1,4,5)trisphosphate and inositol-(1,3,4,5)tetrakisphosphate to their receptor sites in rat cortical membranes. *Neurosci Lett. 191*, 31-34.

[52] Shamir, A., Ebstein, R. P., Nemanov, L., Zohar, A., Belmaker, R. H. & Agam. G. (1998). Inositol monophosphatase in immortalized lymphoblastoid cell lines indicates susceptibility to bipolar disorder and response to lithium therapy. *Mol Psychiatry. 3*, 481-482.

[53] Shafer, T. J., Nostrandt, A. C., Tilson, H. A. & Mundy, W. R. (1994). Mechanisms underlying AlCl3 inhibition of agonist-stimulated inositol phosphate accumulation. Role of calcium, G-proteins, phospholipase C and protein kinase C. *Biochem Pharmacol. 47*, 1417-1425.

[54] Berridge, M. J. (2012). Calcium signalling remodelling and disease. *Biochem Soc Trans. 40*, 297-309.

[55] Berridge, M. J. (1994). The biology and medicine of calcium signalling. *Mol Cell Endocrinol. 98*, 119-124.

[56] Berridge, M. J. (2013). Dysregulation of neural calcium signaling in Alzheimer disease, bipolar disorder and schizophrenia. *Prion. 7*, 2-13.

[57] Mattson, M. P. (2010). ER calcium and Alzheimer's disease, in a state of flux. *Sci Signal. 3*:e10.

[58] Lu, S., Huang, W., Li, X., Huang, Z., Liu, X., Chen, Y., Shi, T. & Zhang. J. (2012). Insights into the role of magnesium triad in myo-inositol monophosphatase, metal mechanism, substrate binding, and lithium therapy. *J Chem. Inf. Model. 52*, 2398-2409.

[59] Newman, P. E. (2000). Alzheimer's disease revisited. *Med Hypotheses. 54*, 774-776.

[60] Barak, Y., Levine, J., Glasman, A., Elizur, A. & Belmaker, R. H. (1996). Inositol treatment of Alzheimer's disease, a double blind, cross-over

placebo controlled trial. *Prog Neuropsychopharmacol. Biol Psychiatry. 20*, 729-735.

[61] Anekonda T. S., Wadsworth, T. I., Sabin, R., Fraler, K., Harris, C., Petriko, B., Ralle, M., Woltjer, R. & Quinn, J. F. J. (2011). Phytic acid as a potential treatment for alzheimer's pathology, evidence from animal and in vitro models. *J. Alzheimers Dis. 23*, 21-35.

In: Inositol
Editors: H. Rocha and M. Cardoso
ISBN: 978-1-62808-520-4

Chapter 7

Design and Synthesis of Biotinylated Inositol Phosphates: Application to the Inositol Phosphate-Protein Binding Analysis

Kensaku Anraku[1], Mikako Fujita[2], Takashi Morii[3], Yasuo Mori[4], Yoshinari Okamoto[5] and Masami Otsuka[5]

[1]Institute of Health Sciences, Kumamoto Health Science University, Izumi-machi, Kita-ku, Kumamoto, Japan
[2]Research Institute for Drug Discovery, School of Pharmacy, Kumamoto University, Oe-honmachi, Chuo-ku, Kumamoto, Japan
[3]Institute of Advanced Energy, Kyoto University, Uji, Kyoto, Japan
[4]Laboratory of Molecular Biology, Department of Synthetic Chemistry and Biological Chemistry, Graduate School of Engineering, Kyoto University, Kyoto, Japan
[5]Department of Bioorganic Medicinal Chemistry, Faculty of Life Sciences, Kumamoto University, Oe-honmachi, Chuo-ku, Kumamoto, Japan

Abstract

Inositol phosphates play important roles as second messengers in intracellular signal transduction. In order to study the relative affinity and specificity in binding inositol phosphates and diverse inositol phosphate-binding proteins by Surface Plasmon Resonance (SPR) analysis and pull-down analysis, biotinylated inositol phosphates, to be immobilized on avidin-based sensor chip or resin, were designed. The synthesis of the biotinylated inositol phosphates was accomplished by assembling the inositol phosphate building block, synthesized starting with optically resolved *myo*-inositol derivatives, and the biotin building block through a phosphate linkage. The inositol phosphate moiety of each biotinylated inositol phosphate showed specific binding to Pleckstrin Homology (PH) domain of PLCδ (phospholipase Cδ), PH domain of Grp1 (general receptor of phosphoinositides 1), and precursor of Gag protein ($Pr55^{Gag}$) of human immunodeficiency virus type 1 (HIV-1), as revealed by SPR analysis and pull-down analysis.

Introduction

D-*myo*-Inositol phosphates (InsP) and phosphatidylinositol phosphates (PtdInsP) are involved in diverse biological phenomena mediated by InsP binding proteins including, e.g., the pleckstrin homology (PH) domain-containing proteins often involved in various intracellular signaling and the $Pr55^{Gag}$ protein that initiates HIV replication. The elucidation of these InsP-binding proteins would not only be a clue to the unknown mechanism of the biological signaling but also lead to drug discovery. These InsP-binding proteins could be best studied by immobilizing InsP on appropriate carriers such as beads or sensor chips. This review deals with synthesis and application of biotinylated inositol phosphates that could be immobilized on various carriers via the biotin-streptavidin interaction.

PH Domain and Calcium Signaling

PH domain, found in over 100 proteins, is a structural module of about 120 amino acids homologous to the two regions in pleckstrin, the major protein kinase C substrate in platelets (Tyers et al. 1988, Haslam et al. 1993, Mayer 1993). PH domain plays significant roles in membrane recruiting of proteins involved in intracellular signaling and the cytoskeleton organization (Kavran et al. 1998, Shaw 1996, Musacchio et al. 1993, Gibson et al. 1994), by binding specific phosphoinositides and their head groups (Harlan et al. 1994, Hyvönen et al. 1995, Lemmon et al. 1995).

Phospholipase C δ (PLCδ) PH domain is a representative D-Ins(1,4,5)P_3-binding motif which recognizes and binds to Ptd-D-Ins(4,5)P_2 with high affinity (Lemmon et al. 1995). D-Ins(1,4,5)P_3 plays a key role in the signaling cascade that links extracellular messengers to intracellular Ca^{2+} mobilization (Berrige 1993). Upon stimulation of a certain receptor, the associated G protein or tyrosine kinase activates a membrane-bound PLC, which hydrolyzes Ptd-D-Ins(4,5)P_2 into D-Ins(1,4,5)P_3 and diacylglycerol (DAG), bifurcating the signaling pathway. The hydrophilic Ins(1,4,5)P_3 diffused in the cytosol activates the receptor of a Ca^{2+} channel on the endoplasmic reticulum, resulting in the release of Ca^{2+} from an internal store.

On the other hand, D-Ins(1,3,4,5)P_4 is formed by the direct phosphorylation of the D-Ins(1,4,5)P_3 by the D-Ins(1,4,5)P_3 3-kinase (Batty et al. 1985, Irvine and Schell 2001). The synergistic involvement of D-Ins(1,3,4,5)P_4 in the D-Ins(1,4,5)P_3-mediated mobilization of Ca^{2+} and the subsequent regulation of the store-operated Ca^{2+}-influx (Kavran et al. 1998, Shaw 1996, Musacchio et al. 1993) still remains as a subject of controversy.

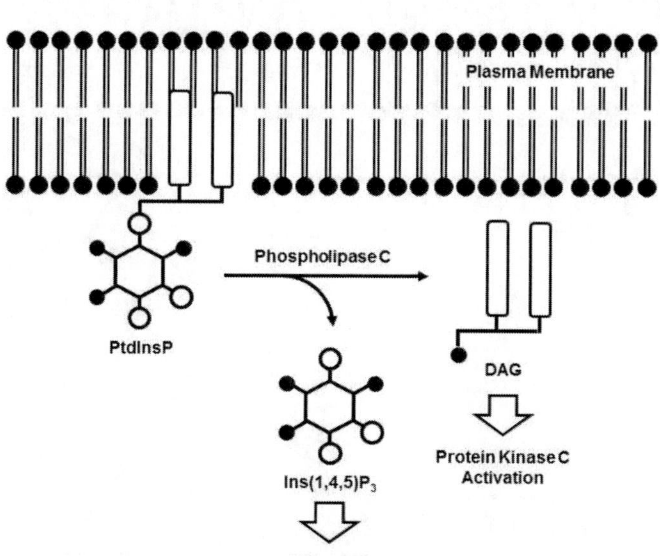

Figure 1. Calcium signaling mediated by D-Ins(1,4,5)P_3 produced by enzyme hydrolysis of membrane Ptd-D-Ins(4,5)P_2.

Gag and HIV-1 Replication

The HIV-1 genome-encoded $Pr55^{Gag}$ protein, the principal structural component required for virus assembly, is known to bind Ptd-D-Ins(4,5)P_2 (Wills and Craven 1991, Freed 1998). Following the ribosomal synthesis, $Pr55^{Gag}$ is directed to the plasma membrane of the host cell, where it is assembled with other components to form premature budding virions. $Pr55^{Gag}$ is composed of four major domains: matrix (MA), capsid (CA), nucleocapsid (NC), and p6, in addition to two spacer peptides: p2 and p1. The N-terminal MA domain plays a critical role in the binding of $Pr55^{Gag}$ to the plasma membrane (Conte and Matthews 1998).

During the membrane targeting, several thousand copies of $Pr55^{Gag}$ colocalize at lipid rafts on the plasma membrane (Nguyen and Hildreth 2000), mediated by the $Pr55^{Gag}$ MA domain N-terminal myristoyl moiety and the basic patch. The myristoyl group attached to the MA N-terminal glycine is flipped and exposed through the multimerization process of $Pr55^{Gag}$ to bind to the plasma membrane; a phenomenon known as the myristoyl switch (Resh 2004, Tang et al. 2004).

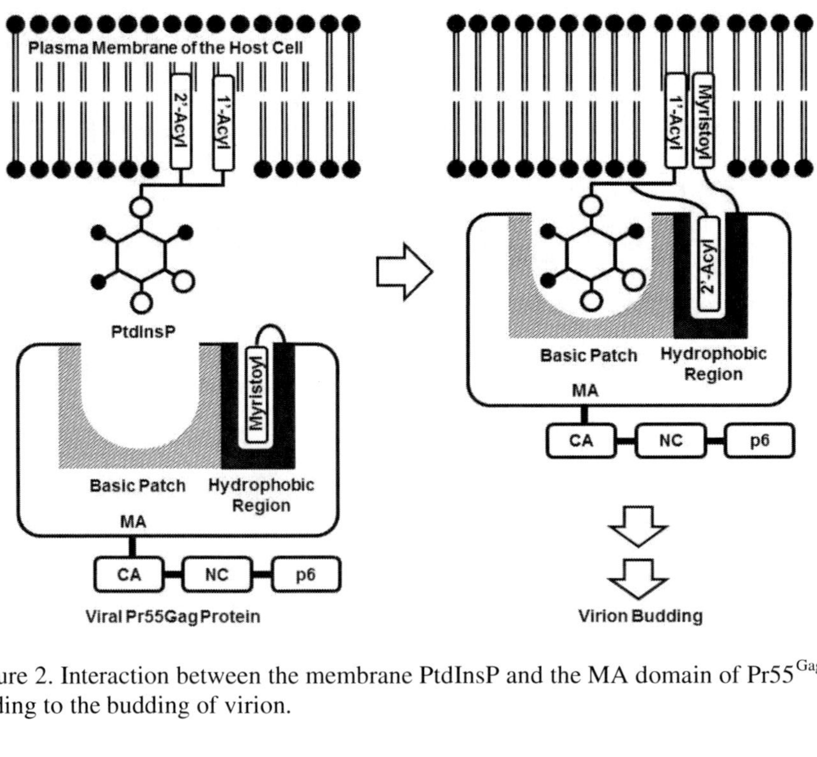

Figure 2. Interaction between the membrane PtdInsP and the MA domain of $Pr55^{Gag}$, leading to the budding of virion.

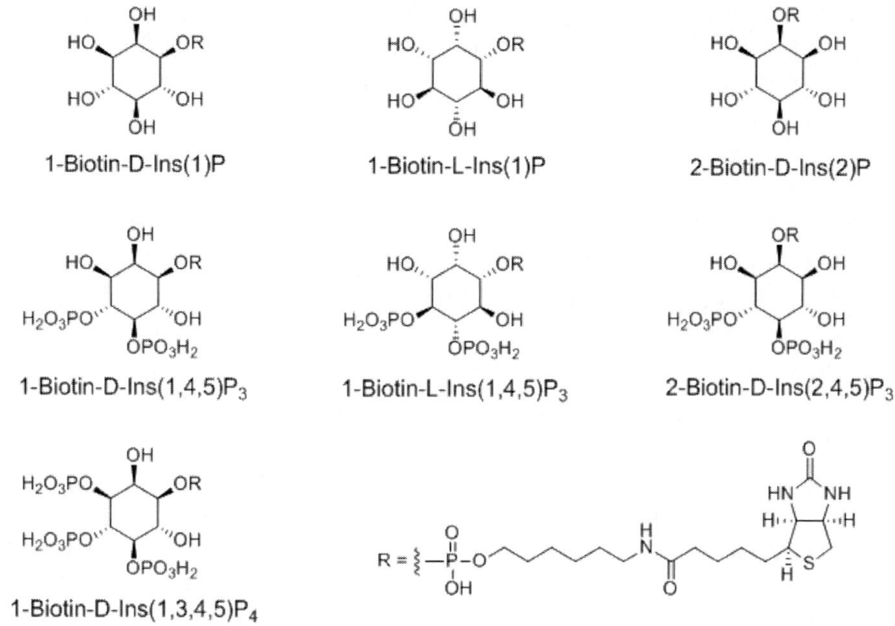

Figure 3. Structures of biotinylated inositol phosphates.

The MA hydrophobic region, in turn, interacts with the 2'-acyl chain of the membrane PtdInsP and the basic patch, on the other hand, accommodates the InsP moiety (Zhou et al. 1994). This mechanism of the MA domain-PtdInsP was supported by the following pioneering studies. Ono et al. identified Ptd-D-Ins(4,5)P_2 as a cellular factor regulating the targeting of $Pr55^{Gag}$ to the plasma membrane (Ono et al. 2004). These findings were supported by NMR analyses by Saad et al. that demonstrated direct interactions between $Pr55^{Gag}$ and Ptd-D-Ins(4,5)P_2, observing a conformational change in $Pr55^{Gag}$ induced by Ptd-D-Ins(4,5)P_2-MA binding that triggers myristate exposure (Saad et al. 2006).

Design of Biotinylated Inositol Phosphates

In order to study the relative affinity and specificity in the binding of InsPs and PH domain or $Pr55^{Gag}$ MA domain, we designed biotinylated inositol phosphates that could be immobilized on avidin-modified beads or sensor chips (Wilchek and Bayer 1990) for the binding analysis.

These include biotin derivatives of inositol mono- and tris- and tetrakisphosphates, *i.e.*, 1-Biotin-D-Ins(1)P, 1-Biotin-L-Ins(1)P, 2-Biotin-D-Ins(2)P, 1-Biotin-D-Ins(1,4,5)P_3, 1-Biotin-L-Ins(1,4,5)P_3, 2-Biotin-D-Ins(2,4,5)P_3, and 1-Biotin-D-Ins(1,3,4,5)P_4 (Anraku et al., 2008, Anraku et al. 2010, Anraku et al. 2011).

Since the X-ray crystal structure of the Ins(1,4,5)P_3-PLCδ PH domain complex (Ferguson et al. 1995) revealed that the inositol and 4,5-phosphate groups of Ins(1,4,5)P_3 are accommodated in the binding pocket, we thought that a biotin-linker could be introduced at the 1-phosphate or 2-phosphate (the latter is an artificially created derivative) of inositol without affecting the PH domain binding.

The X-ray crystal structure of the Ins(1,3,4,5)P_4-Grp1 PH domain complex (Lietzke et al. 2000) revealed that the phosphate groups at the 3-, 4-, and 5-positions of Ins(1,3,4,5)P_4 are accommodated in the binding pocket of the Grp1 PH domain, while 1-phosphate lies out of the binding pocket. Thus, introduction of a biotin-linker at the 1-phosphate of Ins(1,3,4,5)P_4 would not affect the complex formation with Grp1PH domain.

Synthesis of Biotinylated Inositol Phosphates

D-Ins(1)P and L-Ins(1)P Fragments

For the synthesis of 2,3,4,5,6-pentaacetyl-D-inositol **4a**, we selected the optically resolved *myo*-inositol derivative (+)-**1** (Billington et al. 1987) as the starting material. Since (+)-**1** has a benzyl group only at the 1-hydroxyl group, the synthesis was straightforward.

Hydrolysis of the cyclohexylidene groups of (+)-**1** with acid provided pentaol **2a**, which was acetylated to give fully protected **3a**. Treatment of **3a** with H_2/Pd-C gave the debenzylated D-Ins(1)P fragment **4a**. The enantiomeric **4b** was prepared starting with (−)-**1** by the same procedure.

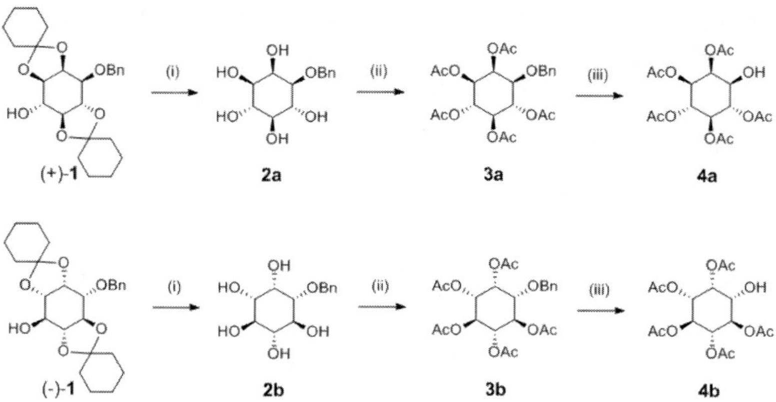

Figure 4. Synthesis of D- and L-Ins(1)P fragments. (i) TFA, MeOH, rt; (ii) Ac_2O, DMAP, Py, rt; iii) H_2/10% Pd-C, CH_3COOH, rt.

D-Ins(1,4,5)P_3 and L-Ins(1,4,5)P_3 Fragments

The synthesis of InsP_3 and InsP_4 was more complicated, necessitating the differentiation of 1- and 2-hydroxyl groups of the inositol. The Nagashima procedure (Nagashima and Ohno 1987) was successfully applied to the case. The stannylene acetal of the 1,2-diol formed was regioselectively alkylated with R^3-X in the presence of cesium fluoride to give 1-*O*-alkylated product.

Thus, the starting material **5a** was prepared by removing the benzyl group of (−)-**1**. Allylation of the diol **5a** provided **6a**, which was further treated with *p*-toluenesulfonic acid and H_2O to give deprotected **7a** in 92% yield (for 2 steps). The *cis*-1,2-diol of **7a** was regioselectively protected by a MPM group by means of the above mentioned Nagashima's dibutyltin oxide procedure (Nagashima and Ohno 1987, Liu and Potter 1997) to give **8a** exclusively in 97% yield.

The introduction of isopropylidene acetal to the 4,5-vicinal alcohol gave **9a** in 85% yield. The subsequent removal of the allyl groups (Corey and Suggs 1973) of **9a** gave **10a** in 67% yield. Acetylation of the 3,6-diol **10a** provided **11a**, which was treated with *p*-toluenesulfonic acid and ethylene glycol to give the 4,5-diol intermediate **12a** in 88% yield (for 2 steps). The subsequent phosphorylation of the 4,5-diol **12a** (Uhlmann and Engels 1986) afforded 4,5-bisphosphate **13a** in 93% yield. Oxidative cleavage of the *p*-methoxybenzyl group with CAN (Johansson and Samuelsson 1984) gave the D-Ins(1,4,5)P_3

fragment **14a** in 71% yield. The enantiomeric **14b** was prepared from (+)-**1** by the same procedure.

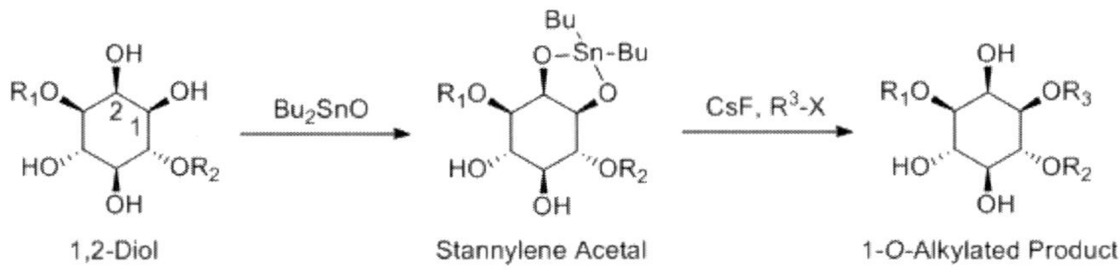

Figure 5. Regioselective alkylation of inositol 1,2-diol via the stannylene acetal formation.

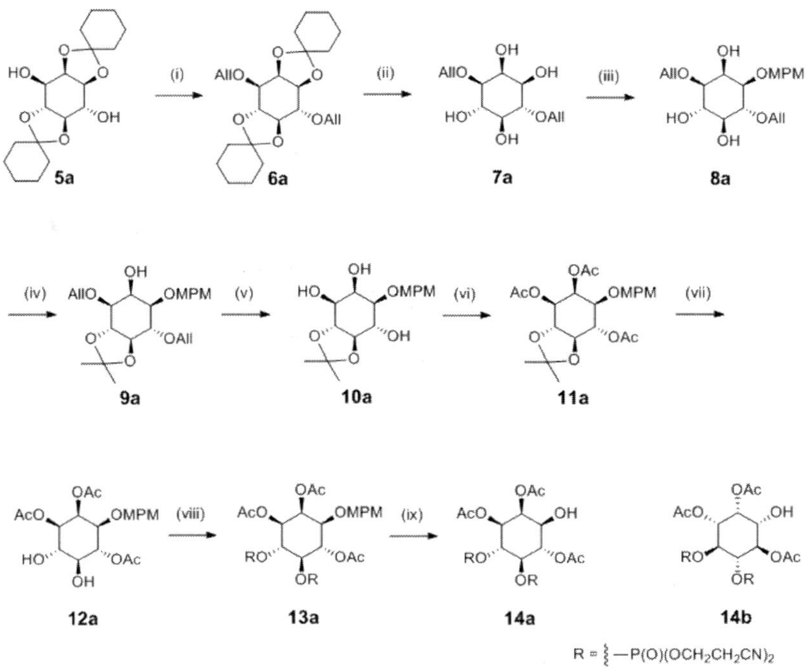

Figure 6. Synthesis of D- and L-Ins(1,4,5)P_3 fragments. (i) Allyl-Br, NaH, DMF, rt; (ii) TsOH, THF–MeOH, reflux; (iii) (a) Bu_2SnO, toluene, reflux, 3 h; (b) CsF, MPM-Cl, DMF, −40°C then rt; (iv) 2-methoxypropene, TsOH, DMF, rt; (v) (a) $(Ph_3P)_3RhCl$, DABCO, EtOH–benzene–H_2O, reflux, 5 h; (b) HgO, $HgCl_2$, acetone–H_2O, rt, 5 min; (vi) Ac_2O, DMAP, Py, rt; (vii) TsOH, ethylene glycol, CH_2Cl_2, rt, 10 min; (viii) (a) bis(2-cyanoethyl)-*N*,*N*-diisopropylphosphoramidite, 1*H*-tetrazole, CH_2Cl_2, rt, 1.5 h; (b) MCPBA, CH_2Cl_2, rt, 5 min; (ix) CAN, CH_3CN–H_2O, rt, 1 h.

D-Ins(2,4,5)P_3 and D-Ins(2)P Fragments

The regioselective allylation (Nagashima and Ohno 1987, Liu and Potter 1997) at the 1-hydroxyl group of **7a** by the Nagashima procedure gave triallyl derivative **15a** in 59% yield. Introduction of isopropylidene acetal to the 4,5-diol **15a** provided **16a**, which was further protected with a MPM group to give the 2-MPM derivative **17a** in 66% yield (for 2 steps). Deprotection of the three allyl groups (Corey and Suggs 1973) of **17a** followed by acetylation gave **19a**. The isopropylidene acetal was removed to afford 4,5-diol **20a** in 42% yield (for 3 steps). Phosphorylation (Uhlmann and Engels 1986) of **20a** gave 4,5-bisphosphate **21a**, which was treated with CAN (Johansson and Samuelsson 1984) to give the D-Ins(2,4,5)P_3 fragment **22a** in 59% yield (for 2 steps).

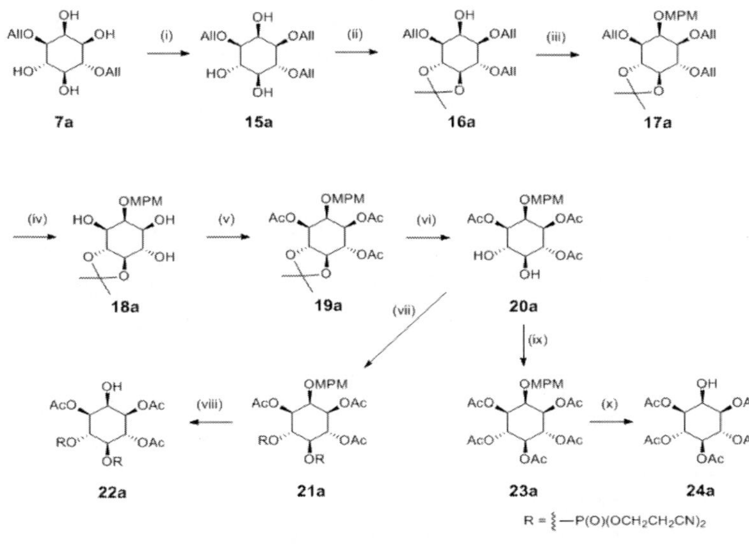

Figure 7. Synthesis of D-Ins(2,4,5)P_3 and D-Ins(2)P fragments. (i) (a) Bu_2SnO, toluene, reflux, 3 h; (b) CsF, Allyl-Br, DMF, −40 °C then rt; (ii) 2-methoxypropene, TsOH, DMF, rt; (iii) MPM-Cl, NaH, DMF, rt; (iv) (a) $(Ph_3P)_3RhCl$, DABCO, EtOH–benzene–H_2O, reflux, 5 h; (b) HgO, $HgCl_2$, acetone–H_2O, rt, 5 min; (v) Ac_2O, DMAP, Py, rt; (vi) TsOH, ethylene glycol, CH_2Cl_2, rt, 10 min; (vii) (a) bis(2-cyanoethyl)-*N*,*N*-diisopropylphosphoramidite, 1*H*-tetrazole, CH_2Cl_2, rt, 1.5 h; (b) MCPBA, CH_2Cl_2, rt, 5 min; (viii) CAN, CH_3CN–H_2O, rt, 1 h; (ix) Ac_2O, DMAP, Py, rt; (x) H_2/10% Pd-C, CH_3COOH, rt.

On the other hand, exhaustive acetylation of **20a** and successive removal of the MPM group of **23a** gave D-Ins(2)P fragment **24a** in 64% yield (for 2 steps).

D-Ins(1,3,4,5)P_4 Fragment

Aiming at the synthesis of 2,5-*O*-diacetyl-1-ol **34**, we started with alcohol (−)-**1** (Billington et al. 1987) Compound **25**, obtained by allylation of the alcohol (−)-**1**, which was treated with *p*-toluenesulfonic acid and H_2O to give deacetalyzed **26** in 67% yield (for two steps). The cis-1,2-diol **26** was regioselectively protected by the Nagashima procedure (Nagashima and Ohno 1987, Liu and Potter 1997) to give regioselectively protected **27** in 88% yield. The introduction of isopropylidene acetal to the 4,5-diol gave **28** in 87% yield. Deprotection of the allyl group of **28** gave **29** in 68% yield (Corey and Suggs 1973).

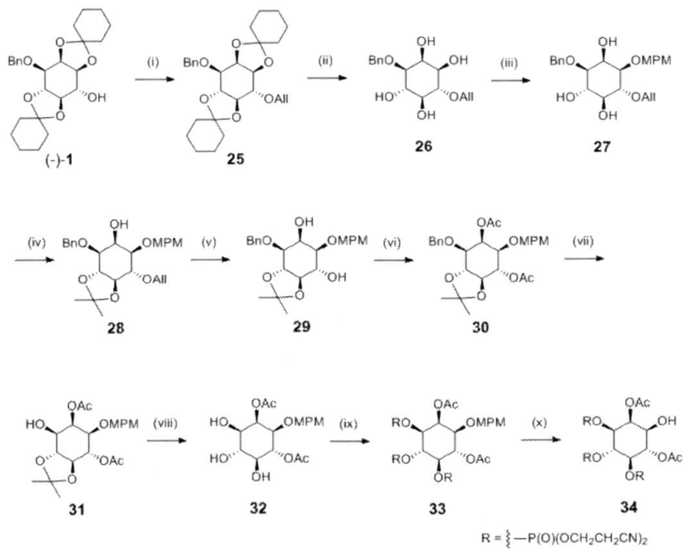

Figure 8. Synthesis of D-Ins(1,3,4,5)P_4 fragment. (i) Allyl-Br, NaH, DMF, rt, 24 h; (ii) TsOH, THF–H_2O, reflux, 3 h; (iii) (a) Bu_2SnO, toluene, reflux, 3 h; (b) CsF, MPM-Cl, DMF, −78 °C then rt, 24 h; (iv) 2-methoxypropene, TsOH, DMF, rt, 24 h; (v) (a) $(Ph_3P)_3RhCl$, DABCO, EtOH–benzene–H_2O, reflux, 5 h; (b) HgO, $HgCl_2$, acetone–H_2O, rt, 5 min; (vi) Ac_2O, DMAP, pyridine, rt, 12 h; (vii) H_2/W-2 Raney-Ni, MeOH, 50 °C, 3 h; (viii) TsOH, ethylene glycol, CH_2Cl_2, rt, 10 min; (ix) (a) bis(2-cyanoethyl)-*N*,*N*-diisopropylphosphoramidite, 1*H*-tetrazole, CH_2Cl_2, rt, 1.5 h; (b) MCPBA, CH_2Cl_2, rt, 5 min; (x) CAN, CH_3CN–H_2O, rt, 1 h.

Acetylation of **29** provided 2,6-diacetate **30** in 91% yield. The selective deprotection of benzyl group of **30** (Oikawa et al. 1984) gave **31** (64% yield), which was treated with *p*-toluenesulfonic acid to give the 3,4,5-triol intermediate **32** in 89% yield. The 3,4,5-triol **32** was converted to the corresponding 3,4,5-trisphosphate **33** by the usual procedure in 51% yield. Oxidative cleavage of MPM group with CAN (Johansson and Samuelsson 1984) gave the D-Ins(1,3,4,5)P_4 fragment **34** in 86% yield.

Synthesis of Biotinylated Inositol Phosphates

The biotin-linker fragment (Pon 1991) was reacted with bifunctional phosphorylating agent (2-cyanoethyl)-*N,N,N,N*-tetraisopropylphosphoramidite (Bannwarth and Trezeciak 1987) and 1*H*-tetrazole to give a rather labile phosphoramidite which was immediately condensed with inositol moiety **14a**, **4a**, **14b**, **4b**, **22a**, **24a** or **34**. Subsequent oxidation of the condensation products with *tert*-BuOOH gave the fully protected biotinylated inositol phosphates **35a** (93% yield), **36a** (78% yield), **35b** (83% yield), **36b** (85% yield), **37** (99% yield), **38** (52% yield), and **39** (<80% yield). The ^{1}H NMR and FABMS data of these compounds showed no oxidation of the biotin sulfide.

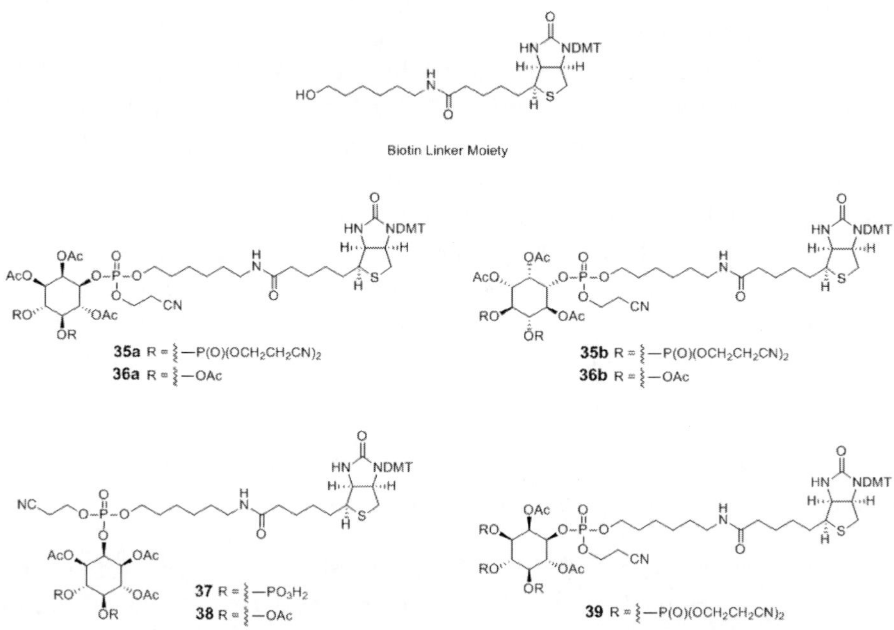

Figure 9. The penultimate precursors for the biotinylated inositol phosphates.

The final stage of the synthesis was carried out as follows. After removal of the trityl group of biotin with acid treatment, all protecting groups were removed in one step by reaction with NH_3 to give water-soluble biotinylated inositol phosphates. The biotinylated products were efficiently purified by anion-exchange chromatography with gradients of ammonium formate as eluent to give 1-Biotin-D-Ins(1,4,5)P_3 (32% yield), 1-Biotin-D-Ins(1)P (30% yield), 1-Biotin-L-Ins(1,4,5)P_3 (35% yield), 1-Biotin-L-Ins(1)P (46% yield), 2-Biotin-D-Ins(2,4,5)P_3 (32% yield), 2-Biotin-D-Ins(2)P (66% yield), and 1-Biotin-D-Ins(1,3,4,5)P_4 (26% yield) (for 2 steps).

Biological Study

PLCδ PH Domain: Surface Plasmon Resonance Analysis

Biological study of the synthetic biotinylated inositol phosphates using streptavidin beads was carried out. The original PLCδ PH domain bound efficiently to the prebound beads of 1-Biotin-D-Ins(1,4,5)P_3. The dissociation constant *K*d of 1-Biotin-D-Ins(1,4,5)P_3 binding of the PLCδ PH domain was 250±20 nM, which was comparable to that of non-tethered D-Ins(1,4,5)P_3 (Lemmon et al. 1995). Further, the Surface Plasmon Resonance (SPR) study of the synthetic biotinylated inositol phosphates against the PLCδ PH domains was carried out using a streptavidin-modified sensor chip. The results are summarized in the Table 1. 1-Biotin-D-Ins(1,4,5)P_3 and the PLCδ PH domain was bound the most tightly (*K*d = 126±23 nM) compared with 1-Biotin-L-Ins(1,4,5)P_3 (*K*d = 262±72 nM) and 2-Biotin-D-Ins(1,4,5)P_3 (*K*d = 303±93). Biotinylated inositol monophosphates did not show discernible binding. These results are consistent with those previously reported by Lemmon et al. (Lemmon et al. 1995).

Grp1 PH Domain: Pull-down Analysis

The biological study of the 1-Biotin-D-Ins(1,3,4,5)P_4 was carried out using streptavidin coated beads. Grp1 PH domain was incubated with various amounts of 1-Biotin-D-Ins(1,3,4,5)P_4 immobilized beads.

Table 1. Dissociation Constant (*K*d) of biotinylated inositol phosphates binding of the PLCδ PH domain (*Lemmon et al. 1995)

	*K*d (nM)	*K*d (nM)*
1-Biotin-D-Ins(1,4,5)P_3	126±23	210±120
1-Biotin-D-Ins(1)P	Not detected	–
1-Biotin-L-Ins(1,4,5)P_3	262±72	8,800
1-Biotin-L-Ins(1)P	Not detected	–
2-Biotin-D-Ins(2,4,5)P_3	303±93	9,600
2-Biotin-D-Ins(2)P	Not detected	>64,000

The binding ratio was calculated by dividing the bound fraction by the sum of bound fraction and unbound supernatant fraction obtained from the SDS–polyacrylamide electrophoretic analyses. An equilibrium dissociation constant *K*d for the complex of 1-Biotin-D-Ins(1,3,4,5)P_4 and Grp1 PH domain was obtained (0.14 ± 0.04 μM) by plotting the binding ratio against 1-Biotin-D-Ins(1,3,4,5)P_4 concentration. According to the previous report (Sakaguchi et al. 2010), the *K*d value for the complex of Grp1 PH domain and Ins(1,3,4,5)P_4 is approximately 0.10 μM, which is in good agreement with the *K*d value in the present pull-down analysis (0.14 μM) despite the presence of the biotin tethered at the 1-phosphate position of inositol. Thus, the result demonstrates that 1-Biotin-D-Ins(1,3,4,5)P_4 successfully reproduces the protein binding characteristics of unmodified D-Ins(1,3,4,5)P_4, and furthermore, the biotin-linker moiety does not perturb the Ins(1,3,4,5)P_4-binding of Grp1 PH domain.

Grp1 PH Domain: Surface Plasmon Resonance Analysis

1-Biotin-D-Ins(1,3,4,5)P_4 was next applied to SPR-based PH domain binding analysis using a streptavidin coated sensor chip. After 1-Biotin-D-Ins(1,3,4,5)P_4 was partially immobilized on the sensor chip, the binding affinity was determined by injecting various concentrations of Grp1 PH domain over the sensor chip. The *K*d was deduced to be 0.3 μM. The reasons for the difference in *K*d values, between SPR-based analysis and Pull-down analysis, could be due to differences in the principle of measurement. To confirm whether the 1-Biotin-D-Ins(1,3,4,5)P_4 retained the binding activity, 1-Biotin-Ins(1,4,5)P_3 and 1-Biotin-L-Ins(1,4,5)P_3 were also immobilized on the sensor chip. Grp1 PH domain barely bound 1-Biotin-D-Ins(1,4,5)P_3 (*K*d >>1.0

μM) and 1-Biotin-L-Ins(1,4,5)P_3 (*K*d >>1.0 μM). Thus, 1-Biotin-D-Ins(1,4,5)P_3 and 1-Biotin-D-Ins(1,3,4,5)P_4 retain the inherent biochemical characteristics for PLCδ domain and Grp1 PH domain, respectively.

Grp1 PH Domain: Docking Study

The mode of interaction of the Grp1 PH domain-D-Ins(1,3,4,5)P_4 and Grp1 PH domain-1-Hexyl-D-Ins(1,3,4,5)P_4 (D-Ins(1,3,4,5)P_4 with hexyl linker at C1 position through a phosphodiester linkage) complexes was examined by a molecular docking study.

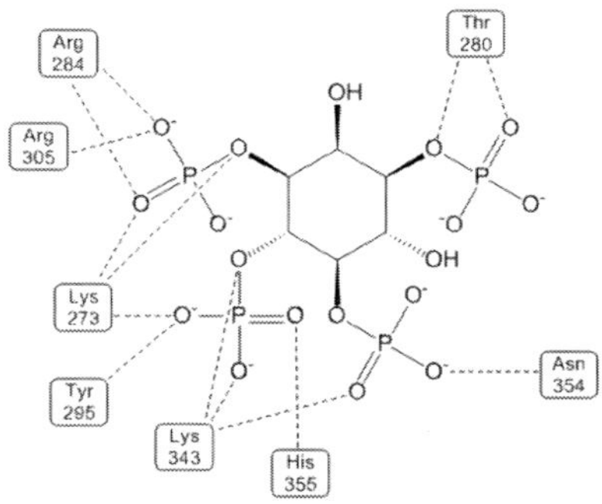

(a)

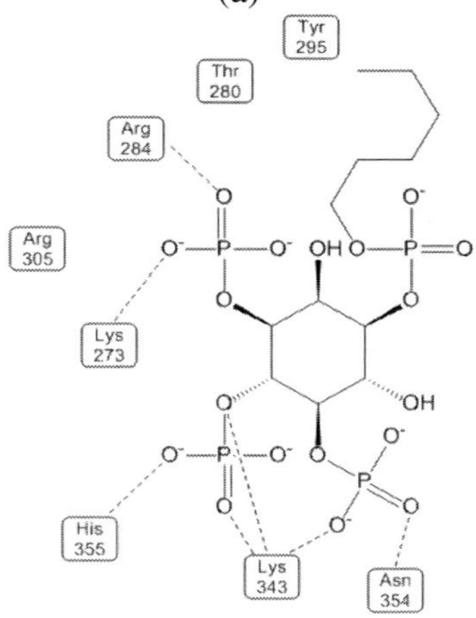

(b)

Figure 10. Docking studies of Grp1 PH domain-unmodified D-Ins(1,3,4,5)P_4 (a) and Grp1 PH domain-1-Hexyl-D-Ins(1,3,4,5)P_4 (b) complexes. The lines show the interaction between amino acids and phosphates within 3.0 Å distance.

Table 2. Interactions between the phosphate/hexyl groups and the Grp1 PH domain

	D-Ins(1,3,4,5)P_4	1-Hexyl-D-Ins(1,3,4,5)P_4
1-Phosphate	Thr280 (2.1, 2.21 Å: OH)	No interaction
3-Phosphate	Lys273 (2.04, 2.66 Å: NH_2) Arg284 (1.84 Å: NH, 1.93 Å: NH) Arg305 (2.19 Å: NH_2)	Lys273 (1.21 Å: NH_2) Arg284 (1.24 Å: NH, 2.39 Å: NH)
4-Phosphate	Lys273 (1.94 Å: NH_2), Tyr295 (1.58 Å: OH) Lys343 (2.01, 2.65, 1.81 Å: NH_2) His355 (1.61 Å: NH)	Lys343 (1.26, 2.63 Å: NH_2) His355 (1.33 Å: NH)
5-Phosphate	Lys343 (1.81 Å: NH_2) Asn354 (1.99 Å: NH_2)	Lys343 (1.2 Å: NH_2) Asn354 (1.81 Å: NH_2)
Hexyl linker		Out of the binding pocket

The structures are shown in Figure 10, wherein the dotted lines indicate the interaction between amino acids and phosphate of D-Ins(1,3,4,5)P_4 (or 1-Hexyl-D-Ins(1,3,4,5)P_4) are shorter than 3.0 Å. Interactions between the phosphate and the hexyl groups and the Grp1 PH domain are summarized in Table 2.

The average of distances between the amino acids and the 1-Hexyl-D-Ins(1,3,4,5)P_4 was shorter than that of unmodified D-Ins(1,3,4,5)P_4 and the numbers of interaction were comparable to that of unmodified D-Ins(1,3,4,5)P_4. In this context, both 1-Biotin-D-Ins(1,3,4,5)P_4 and unmodified D-Ins(1,3,4,5)P_4 were reasonably accommodated in the same binding pocket of the Grp1 PH domain. The 1-phosphate group of 1-Hexyl-D-Ins(1,3,4,5)P_4 did not interact with Thr280 and the hexyl group located outside the binding pocket of Grp1 PH domain. Thus, the length of the linker (C_6) is suitable for the complex formation of 1-Biotin-D-Ins(1,3,4,5)P_4 and the Grp1 PH domain.

$Pr55^{Gag}$/MA: Surface Plasmon Resonance Analysis

1-Biotin-D-Ins(1,3,4,5)P_4 was immobilized to the streptavidin sensor chip in the SPR system through the biotin-streptavidin interaction (Anraku et al. 2011). MA and full length $Pr55^{Gag}$ were expressed by 293T cells transfected with pEF/Myc-His A-based vector and pSG5-based vector, respectively (Anraku et al. 2010). The MA protein thus produced was myristoylated as confirmed by MALDI-TOF-MS. The *K*d of Pr55^{Gag}/MA-phosphoinositide complexes were calculated using a competition assay. The competitors used in this study are shown in Figure 11.

The *K*d values of Pr55^{Gag} and MA, complexed with various phosphoinositides, are shown in Table 3. To more fully define these interactions, we examined the binding affinity between Pr55^{Gag} and D-Ins(1,4,5)P_3, which has divalent phosphate groups at the 1,4,5-position of the inositol head group and lacks an acyl tail. We found that D-Ins(1,4,5)P_3 binds Pr55^{Gag} very weakly (*K*d = 2170 μM). We then determined the binding affinity between Pr55^{Gag} and di-C_8-Ptd-D-Ins, which is composed of a glycerol tail containing saturated C_6 acyl chains and inositol devoid of divalent phosphate groups. Unexpectedly, di-C_8-Ptd-D-Ins bound Pr55^{Gag} with an affinity 12 times greater (*K*d = 186 μM) than that of D-Ins(1,4,5)P_3 (*K*d = 2170 μM). The *K*d for the interaction of Pr55^{Gag} and di-C_8-Ptd-D-Ins(4,5)P_2, which has divalent phosphate groups at the 4,5-position of the inositol head group and saturated C_6 acyl chains, was 47.4 μM. McLaughlin and Aderem suggested that the myristoyl group of Pr55^{Gag} binds to lipid bilayers with a *K*d of approximately 100 μM (McLaughlin and Aderem 1995); thus, the binding affinity we determined for di-C_8-Ptd-D-Ins(4,5)P_2 (*K*d = 47.4 μM) is sufficient for membrane binding of Pr55^{Gag}.

In addition, the binding affinity of di-C_8-Ptd-D-Ins(4,5)P_2 was 46 and 4 times greater than that of D-Ins(1,4,5)P_3 and di-C_8-Ptd-D-Ins, respectively. The same tendency was observed in the case of complexes of other phosphoinositide derivatives and MA.

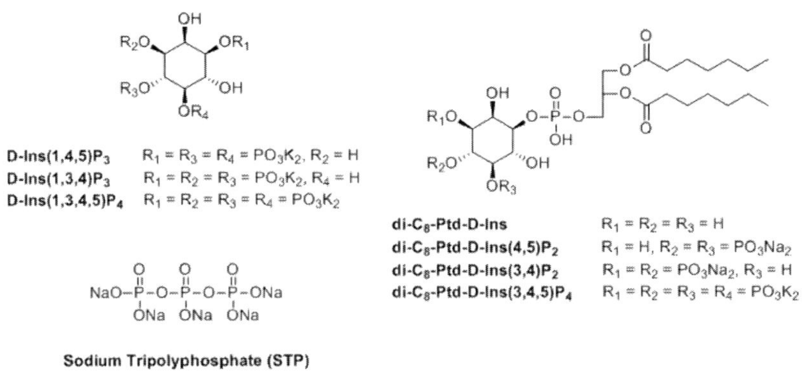

Figure 11. Structures of competitors used in the competition assay.

Table 3. K*d* Values for Pr55^{Gag} or MA Complexed with Phosphoinositide-Related Compounds

Proteins	Competitors	*K*d (μM)
Pr55^{Gag}	InsPs	
	D-Ins(1,4,5)P_3	2,170±260
	D-Ins(1,3,4)P_3	579±48
	D-Ins(1,3,4,5)P_4	332±58
	Phosphate	
	STP	1,810±240
	PtdInsPs	
	di-C_8-Ptd-D-Ins	186±22
	di-C_8-Ptd-D-Ins(4,5)P_2	47.4±6.5
	di-C_8-Ptd-D-Ins(3,4)P_2	54.6±4.8
	di-C_8-Ptd-D-Ins(3,4,5)P_3	19.3±2.9
MA	InsPs	
	D-Ins(1,4,5)P_3	568±52
	D-Ins(1,3,4,5)P_4	526±57
	PtdInsPs	
	di-C_8-Ptd-D-Ins	178±29
	di-C_8-Ptd-D-Ins(4,5)P_2	5.64±1.43
	di-C_8-Ptd-D-Ins(3,4)P_2	2.51±1.56
	di-C_8-Ptd-D-Ins(3,4,5)P_3	6.02±0.97
	di-C_4-Ptd-D-Ins(4,5)P_2	86.6±2.2

These data suggest that both phosphate groups and acyl chains are essential for tight binding between Pr55^{Gag} and phosphoinositides and that the presence of the acyl chains is more important than the inositol divalent phosphate groups.

Conclusion

Biotinylated inositol phosphates were designed to be immobilized on beads or SPR sensor chips. These were efficiently synthesized via routes utilizing the regioselective alkylation of 1,2-diol. Each biotinylated inositol phosphate synthesized showed specific binding to the PLCδ PH domain, Grp1 PH domain, and Pr55^{Gag} protein as revealed by SPR analysis and pull-down analysis. Biotinylated inositol phosphates would be applicable to function analysis of diverse inositol phosphate-binding proteins.

References

Anraku, K.; Inoue, T.; Sugimoto, K.; Morii, T.; Mori, Y.; Okamoto, Y.; Otsuka, M. Design and synthesis of biotinylated inositol phosphates relevant to the biotin-avidin techniques. *Org. Biomol. Chem.*, 2008, 6, 1822-1830.

Anraku, K.; Fukuda, R.; Takamune, N.; Misumi, S.; Okamoto, Y.; Otsuka, M.; Fujita, M. Highly sensitive analysis of interaction between HIV-1 Gag and phosphoinositide derivatives based on surface plasmon resonance. *Biochemistry*, 2010, 49, 5109-5116.

Anraku, K.; Inoue, T.; Sugimoto, K.; Kudo, K.; Okamoto, Y.; Morii, T.; Mori, Y.; Otsuka, M. Design and synthesis of biotinylated inositol 1,3,4,5-tetrakisphosphate targeting Grp1 pleckstrin homology domain. *Bioorg. Med. Chem.*, 2011, 19, 6833-6841.

Bannwarth, W.; Trzeciak, A. A simple and effective chemical phosphorylation procedure for biomolecules. *Helv. Chim. Acta*, 1987, 70, 175-186.

Batty, I. R.; Nahorski, S. R.; Irvine, R. F. Rapid formation of inositol 1,3,4,5-tetrakisphosphate following muscarinic receptor stimulation of rat cerebral cortical slices. *Biochem. J.*, 1985, 232, 211-215.

Berridge, M. J. Inositol trisphosphate and calcium signalling. *Nature*, 1993, 361, 315-325.

Billington, D. C.; Baker, R.; Kulagowski, J. J.; Mawer, I. M. Synthesis of *myo*-inositol 1-phosphate and 4-phosphate, and of their individual enantiomers. *J. Chem. Soc., Chem. Commun.*, 1987, 4, 314-316.

Classon, B.; Garegg, P.; Samuelsson, B. The *p*-methoxybenzyl group as protective group of the anomeric centre. Selective conversions of hydroxyl groups into bromo groups in *p*-methoxybenzyl 2-deoxy-2-phthalimido-*β*-D-glucoside. *Acta Chem. Scand.*, 1984, 38b, 419-422.

Conte, M. R.; Matthews, S. Retroviral matrix proteins: a structural perspective. *Virology*, 1998, 246, 191-198.

Corey, E. J.; Suggs, W. Selective cleavage of allyl ethers under mild conditions by transition metal reagents. *J. Org. Chem.*, 1973, 38, 3224.

Ferguson, K. M.; Lemmon, M. A.; Sigler, P. B.; Schlessinger, J. Structure of the high affinity complex of inositol trisphosphate with a phospholipase C pleckstrin homology domain. *Cell*, 1995, 83, 1037-1046.

Freed, E. O. HIV-1 gag proteins: diverse functions in the virus life cycle. *Virology*, 1998, 251, 1-15.

Gibson, T.; Hyvönen, M.; Musacchio, A.; Saraste, M.; Birney, E. PH domain: the first anniversary. *Trends Biochem. Sci.*, 1994, 19, 349-353.

Haslam, R. J.; Koide, H. B.; Hemmings, B. A. Pleckstrin domain homology. *Nature*, 1993, 363, 309-310.

Harlan, J. E.; Hajduk, P. J.; Yoon, H. S.; Fesik, S. W. Pleckstrin homology domains bind to phosphatidylinositol-4,5-bisphosphate. *Nature*, 1994, 371, 168-170.

Hyvönen, M.; Macias, M. J.; Nilges, M.; Oschkinat, H.; Saraste, M.; Wilmanns, M. Structure of the binding site for inositol phosphates in a PH domain. *EMBO J.*, 1995, 14, 4676-4685.

Irvine, R. F.; Schell, M. J. Back in the water: the return of the inositol phosphates. *Nat. Rev. Mol. Cell Biol.*, 2001, 2, 327-338.

Johansson, R.; Samuelsson, B. Regioselective reductive ring-opening of 4-methoxybenzylidene acetals of hexopyranosides. Access to a novel protecting-group strategy. Part 1. *J. Chem. Soc., Perkin Trans. 1*, 1984, 2371-2374.

Kavran, J. M.; Klein, D. E.; Lee, A.; Falasca, M.; Isakoff, S. J.; Skolnik, E. Y.; Lemmon, M. A. Specificity and promiscuity in phosphoinositide binding by pleckstrin homology domains. *J. Biol. Chem.*, 1998, 273, 30497-30508.

Lemmon, M. A.; Ferguson, K. M.; O'Brien, R. ; Sigler, P. B.; Schlessinger, J. Specific and high-affinity binding of inositol phosphates to an isolated pleckstrin homology domain. *Proc. Natl. Acad. Sci. U. S. A.*, 1995, 92, 10472-10476.

Lietzke, S. E. ; Bose, S.; Cronin, T.; Klarlund, J.; Chawla, A.; Czech, M. P.; Lambright, D. G. Structural basis of 3-phosphoinositide recognition by pleckstrin homology domains. *Mol. Cell*, 2000, 6, 385-394.

Liu, C.; Potter, B. L. Synthesis of 3-position-modified analogues of *myo*-inositol 1,4,5-trisphosphate, tools for investigation of the polyphosphoinositide pathway of cellular signaling. *J. Org. Chem.*, 1997, 62, 8335-8340.

Mayer, B. J.; Ren, R.; Clark, K. L.; Baltimore, D. A putative modular domain present in diverse signaling proteins. *Cell*, 1993, 73, 629-630.

McLaughlin, S.; Aderem, A. The myristoyl-electrostatic switch: a modulator of reversible protein-membrane interactions. *Trends Biochem. Sci.*, 1995, 20, 272-276.

Musacchio, A.; Gibson, T.; Rice, P.; Thompson, J.; Saraste, M. The PH domain: a common piece in the structural patchwork of signalling proteins. *Trends Biochem. Sci.*, 1993, 19, 343-348.

Nagashima, N.; Ohno, M. An efficient *O*-monoalkylation of dimethyl L-tartrate via *O*-stannylane acetal with alkyl halides in the presence of cesium fluoride. *Chem. Lett.*, 1987, 141-144.

Nguyen, D. H.; Hildreth, J. E. Evidence for budding of human immunodeficiency virus type 1 selectively from glycolipid-enriched membrane lipid rafts. *J. Virol.*, 2000, 74, 3264-3272.

Oikawa, Y.; Tanaka, T.; Horita, K.; Yonemitsu, O. Selective hydrogenolysis of the benzyl protecting group for hydroxy function with raney nickel in the presence of the MPM (4-methoxybenzyl) and DMPM (3,4-dimethoxybenzyl) protecting groups. *Tetrahedron Lett.*, 1984, 25, 5397-5400.

Ono, A. ; Ablan, S. D.; Lockett, S. J.; Nagashima, K.; Freed, E. O. Phosphatidylinositol (4,5) bisphosphate regulates HIV-1 Gag targeting to the plasma membrane. Proc. *Natl. Acad. Sci. U.S.A.*, 2004, 101, 14889-14894.

Pon, R. T. A long chain biotin phosphoramidite reagent for the automated synthesis of 5'-biotinylated oligonucleotide. *Tetrahedron Lett.*, 1991, 32, 1715-1718.

Resh, M. D. A myristoyl switch regulates membrane binding of HIV-1 Gag. *Proc. Natl. Acad. Sci. U.S.A.*, 2004, 101, 417-418.

Saad, J. S. ; Miller, J.; Tai, J.; Kim, A.; Ghanam, R. H.; Summers, M. F. Structural basis for targeting HIV-1 Gag proteins to the plasma membrane for virus assembly. *Proc. Natl. Acad. Sci. U.S.A.*, 2006, 103, 11364-11369.

Sakaguchi, R.; Tainaka, K.; Shimada, N.; Nakano, S.; Inoue, M.; Kiyonaka, S.; Mori, Y.; Morii, T. An in vivo fluorescent sensor reveals intracellular Ins(1,3,4,5)P_4 dynamics in single cells. *Angew. Chem., Int. Ed. Engl.*, 2010, 49, 2150-2153.

Shaw, G. The pleckstrin homology domain: an intriguing multifunctional protein module. *BioEssays*, 1996, 18, 35-46.

Tang, C.; Loeliger, E.; Luncsford, P.; Kinde, I.; Beckett, D.; Summers, M. F. Entropic switch regulates myristate exposure in the HIV-1 matrix protein. *Proc.Natl. Acad. Sci.U.S.A.*, 2004, 101, 517-522.

Tyers, M.; Rachubinski, R. A.; Stewart, M. I.; Varrichio, A. M.; Shorr, R. G. L.; Haslam, R. J.; Harley, C. B. Molecular cloning and expression of the major protein kinase C substrate of platelets. *Nature*, 1988, 333, 470-473.

Uhlmann, E.; Engels, J. Chemical 5'-phosphorylation of oligonucleotides valuable in automated DNA synthesis. *Tetrahedron Lett.*, 1986, 27, 1023-1026.

Wilchek, M.; Bayer, E. A. Biotin-containing reagents. *Methods Enzymol.*, 1990, 184, 123-138.

Wills, J. W.; Craven, R. C. Form, function, and use of retroviral Gag proteins. *AIDS*, 1991, 5, 639-654.

Zhou, W.; Parent, L. J.; Wills, J. W.; Resh, M. D. Identification of a membrane-binding domain within the amino-terminal region of human immunodeficiency virus type 1 Gag protein which interacts with acidic phospholipids. *J. Virol.*, 1994, 68, 2556-2569.

In: Inositol
Editors: H. Rocha and M. Cardoso
ISBN: 978-1-62808-520-4

Chapter 8

Health Benefits of Inositol and Inositol Hexaphosphate

Tzi Bun Ng and Charlene Cheuk Wing Ng*
School of Biomedical Sciences, Faculty of Medicine, The Chinese University of Hong Kong, New Territories, Hong Kong, China

Abstract

Inositol hexaphosphate (InsP6, IP6, phytic acid) is found in abundance in beans and cereals. InsP6 occurs at much lower levels in most mammalian cells, in which they regulate cellular functions. Altered inositol production has been reported from patients with diabetes mellitus, chronic renal failure, galactosemia, and multiple sclerosis.IP6 has drawn much attention due to its activity in cancer prevention and control of experimental tumor growth, progression, and metastasis. InsP6 increases differentiation of malignant cells which may then revert to the normal phenotype. IP6 treatment of all the cell lines tested so far demonstrates that it is cytostatic and not cytotoxic. InsP6 inhibits growth and induces terminal differentiation of HT-29 human colon cancer cells. Studies of the expression of tumor suppressor gene demonstrate up-regulation of wild type p53 and down-regulation of the mutant form. Its efficacy in treating central nervous system disorders including Alzheimer's disease, depression, panic disorder, and obsessive-compulsive disorder has been reported. Its usefulness as an analgesic, in

* Email: b021770@mailserv.cuhk.edu.hk, fax: 852 26035123, phone: 852 39436872.

pediatric respiratory depression syndrome,to mitigate adverse effects associated with lithium treatment, and in preventing neural tube defects in embryonic mice has been documented. IP(6) exhibits immunoenhancing and hypocholesterolemic activities, minimizes pathological calcification and kidney stone formation, and lowers pathological platelet activity.However,the risk of inducing uterine contractions restricts its usefulness in pregnancy.

Keywords: Inositol, phytic acid, anticancer, immunoregulatory, calcification, osteoclatogenesis, antioxidant

Introduction

Phytic acid (IP6) is an essential component of high fiber diet. It is physiologically present in the human large gut at concentrations reaching 4 mM.The intent of this article is to review the various activities of inositol, IP6 and related compounds such as phosphatidylinositol which may be exploited for therapeutic purposes.

A diversity of functions has been found for phosphatidylinositol in biological membranes comprising mediation of cellular responses to exogenous stimuli, neuro- transmission, and regulation of enzymatic activity through specific interactions with a variety of proteins. Inositol deficiency in animals ensues in hepatic triglyceride accumulation, intestinal lipodystrophy, and other derangements. Altered inositol metabolism takes place in patients with diseases such as chronic renal failure, diabetes mellitus, galactosemia, and multiple sclerosis. A moderate rise in plasma and nerve inositol levels brought about by dietary supplementation has been proposed for treating diabetic neuropathy, although overly elevated levels, such as those detected in uremic patients, may be neurotoxic (Holub, 1982).

Nutritional and biochemical research have disclosed inositol as a vital dietary and cellular constituent. The functions of phosphatidylinositol in biological membranes consist of control of cellular responses to exogenous stimuli and/or neurotransmission and mediation of enzyme activity via interactions with specific proteins. The efficacy of inositol in the treatment of central nervous system disorders such as Alzheimer's disease, depression, panic disorder, and obsessive-compulsive disorders, and in other diseases including pediatric respiratory depression syndrome, and as potential treatment to alleviate some negative effects associated with lithium therapy.Its usefulness as an analgesic has been evaluated. The use of inositol in pregnant women remains controversial. Although inositol can prevent neural tube defects in mouse embryos, the risk of induction of uterine contractions restricts its application in pregnancy (Colodny and Hoffman, 1998).

Inhibition of Osteoclastogenesis

It has been suggested that IP6 exerts effects on bone resorption analogous to those of bisphosphonates.It suppresses osteoclastogenesis in RAW 264.7 murine monocyte/macrophage cells induced by RANKL, without adversely affecting cell proliferation or cell viability.It also displays an inhibitory action on osteoclastogenesis of human peripheral blood mononuclear cells and their resorption activity both, when given to undifferentiated and to mature osteoclasts.Hence, IP6 represents a new class of selective inhibitor of osteoclasts and may find application in the therapy of osteoporosis (Arriero Mdel et al., 2012).

Immunoregulatory Action on Chickens

The effects of phytate (IP6) and *Escherichia coli*-derived phytase on broiler performance and immunity (both cellular and humoral) were investigated by Liu et al. (2008b),employing female broilers fed nutritionally marginal diets. It was disclosed that phytase augmented feed intake, body weight, and feed conversion ratio.In contrast, high dietary phytate dereased the aforementioned parameters. There was no interaction between phytate and phytase regarding the parameters. Dietary phytase elevated the percentages of erythrocyte rosette-forming cells and erythrocyte-antibody complement cells, and the percentages of CD4(+)CD8(+) T lymphocyte subsets, without changing the CD4(+) toCD8(+)ratio. The levels of intestinal secretory IgA were elevated by phytase. Production of antibodies against Newcastle disease virus vaccine was heightened in the high-phytate diets with phytase added. The data indicate that phytase inclusion in nutritionally marginal diets can increase the number of lymphocytes and the seral and mucosal antibodies, signifying that both phytate and phytase are important to gastrointestinal health and immune competence (Liu et al., 2008b).

Role in Immune Response Through Modulation of Secretion of Proinflammatory Cytokines and Chemokines

The p38 pathway signal transduction pathways is implicated in inflammatory responses. Exposure of Caco-2 cells to IP6 for 3 h brought about an attenuation in p38 mitogen-activated protein kinase (MAP kinase) expression in both unstimulated cells and IL-1beta-stimulated cells. Exposure of Caco-2 cells to anisomycin, a p38 MAP kinase activator enhanced secretion of IL-8 whereas anisomycin pretreatment before addition of IP6 reduced secretion of IL-8 compared to cells exposed to only anisomycin. Thus p38 MAPK is a possible IP6 molecular target in the intestinal epithelial cells. The suppressive effect of IP6 on IL-8 secretion by Caco-2 cells is probably mediated by lowering activity of MAP kinase (Wawszczyk et al., 2012).

Immunoregulatory Effects on Intestinal Epithelium

Intestinal epithelial cells play a crucial part in mucosal immune and inflammatory reactions through the expression and secretion of proinflammatory cytokines including interleukin-6 (IL-6) and interleukin-8 (IL-8). The expression of IL-6 and IL-8 is under the regulation of nuclear factor KB (NF-kappaB). IP6 can modulate immune functions of intestinal epithelium through regulation of proinflammatory cytokines secretion. Treatment with IP6 at 1 and 2.5 mM concentrations for 3, 6 and 12 h was devoid of any effect on gene transcription of p65 subunit of NF-kappaB and modulated gene expression of p50 subunit of NF-kappaB and IkappaBalpha (NF-kappaB inhibitor) in IL-1beta stimulated Caco-2 cells. IP6 brought about a pronounced decline in both IL-6 (at 3 and 6 h) and IL-8 expression (3 h). Thus IP6 may elicit immunoregulating actions on intestinal epithelium by affecting the transcription of genes encoding the p50 subunit of, the NF-kappaB inhibitor IkappaBalpha and the proinflammatory cytokines IL-6 and IL-8 (Wawszczyk et al., 2012).

Enhancement of Gastrointestinal Absorption of Anthocyanins

IP6 is known to promote gastrointestinal absorption of blackcurrant anthocyanins. (Matsumoto et al., 2007).

Anticancer Activity

In vitro anticancer efficacy of IP6 has been observed in many human, mouse and rat prostate cancer cells. Oral feeding of IP6 inhibits human PCA xenograft growth in nude mice without toxicity.In a recently completed pilot study, we observed similar preventive effects of IP6 on prostate tumorigenesis in the TRAMP model. Mechanistic studies indicate that IP6 targets mitogenic and survival signaling, as well as cell cycle progression, in PCA cells. IP6 is also shown to target molecular events associated with angiogenesis. Moreover, IP6 has pleiotropic molecular targets for its overall efficacy against PCA and, therefore, could be a suitable candidate agent for preventive intervention of this malignancy in humans (Singh and Agarwal, 2005).

IP6 exerted inhibitory effects against pancreatic cancer in vitro. IP6 reduced growth of melanoma HTB68 cells and VEGF production and increased late apoptosis in melanoma (Rizvi et al., 2006).

IP6 has arrested considerable attention for its ability to prevent cancer and control experimental tumor growth, progression, and metastasis. Exogenous IP6 is quickly internalized into cells and undergoes dephosphorylation to lower inositol phosphates, further influencing signal transduction pathways leading to cell cycle arrest.Other than reducing cell proliferation, IP6 also induces malignant cell differentiation. Heightened immunity and antioxidant properties also contribute to tumor cell destruction.IP6 potentiates the anticancer effect of conventional chemotherapy, controls cancer metastases, and betters the quality of life. Since it is present in the diet and readily absorbed from the gastrointestinal tract, and safe, IP6 is useful for cancer prevention and therapy (Vucenik and Shamsuddin, 2006).

IP6 inhibited growth in liver, ovary, breast and cancer cells with 50% inhibition concentration (IC50) values of 1.66, 3.45, 3.78 and mM, respectively, but did not affect the normal 3T3 cell line. IP6 exerted antioxidant activity when assayed using the ferric thiocyanate, thiobarbituric acid, ferric reducing antioxidant power and beta-carotene bleaching tests but antioxidant activity could not be attributed to the scavenging of free radical species as determined by the DPPH radical scavenging assay.The antioxidant activity of IP6 may contribute to its anticancer action (Norhaizan et al., 2011).

Khatiwada et al. (2011) demonstrated the inhibitory effects of combinations of 1% and 2% green tea catechins, phytic acid, and inositol in diminishing the incidence of azoxymethane-induced colon tumors and the number of tumors per tumor-bearing rat in male rats (Khatiwada et al., 2011).

Transforming growth factors-beta isoforms (TGF-beta1, 2, and 3) are cytokines that potently suppress uncontrolled cell proliferation. IP6 enhanced transcriptional activity of genes of all three isoforms and receptors of TbetaRI, TbetaRII TbetaRIII in Caco cells (Kapral et al., 2013).

Antioxidant and Carbohydrase Inhibiting Activities

Cooking and roasting increased the antioxidant (DPPH radical scavenging capacity) and health relevant functionality (α-amylase inhibition activity and α-glucosidase inhibition) of phytic acid extracts produced from Kenyan local vegetables and grains (Kunyanga et al., 2011).

Effect on Neurological Diseases

Alzheimer's disease (AD) causes progressive, age-dependent cortical and hippocampal dysfunction leading to abnormal intellectual capacity and memory. In MC65 cells, phytic acid treatment for 48-72 hours completely protected against amyloid precursor protein-C-terminal fragment-induced cytotoxicity by reducing the levels of amyloid β oligomers, intracellular calcium, hydrogen peroxide, and superoxide, and augmented the expression of autophagy (beclin-1) protein. Wild type and twelve-month-old Tg2576 mice were treated for 6 months with 2% phytic acid or vehicle. Modest effects of phytic acid on the expression of amyloid-β precursor protein AβPP trafficking-associated protein AP180, autophagy-associated proteins (beclin-1, LC3B), sirtuin 1, the ratio of phosphorylated AMP-activated protein kinase (PAMPK) to AMPK, soluble amyloid-β1-40, and insoluble amyloid-β1-42 were observed. These results suggest that phytic acid may be useful for treatment of AD (Anekonda et al., 2011).

According to Levine (1997), the level of inositol in the cerebrospinal fluid is reduced in depression. Inositol has therapeutic effects in depression, panic and obsessive-compulsive disorder (OCD), but is not beneficial in schizophrenia, Alzheimer's ADDH, autism or ECT-induced cognitive impairment.

A previous study suggested that inositol alone is efficacious in OCD. Ten OCD patients each took 18 g inositol or placebo for 6 wk in addition to ongoing Current serotonin reuptake inhibitor treatment in a double-blind randomized cross-over design. Weekly assessments included the Yale-Brown Obsessive-Compulsive Scale (YBOCS) and Hamilton Depression and Anxiety scales. No significant difference was found between the two treatment phases (Fux et al., 1999).

Larzelere et al. (2010) are of the opinion that there is inadequate evidence for a beneficial action of inositol on depression but preliminary evidence supports the use of inositol for the treatment of anxiety.

Inhibition of Development of Dental Calculus Deposits

Dental calculi result from saliva supersaturation regarding calcium phosphates. This dental plaque mineralization can be procrastinated by crystallization inhibitors like bisphosphonates and pyrophosphate. Phytate, added to a mouthwash solution, was very effective in preventing dental calculus formation (Grases et al., 2009).

Inhibition of Soft Tissue Calcification

Calcification was significantly suppressed in phytate-treated rats after 21 days of treatment compared with untreated rats. Calcified deposits were detected as early as the third day following induction of calcinosis by subcutaneous potassium permanganate injection at two sites on either side of the interscapular region. Inflammation with accumulation of eosinophils, lymphocytes and macrophages was indiscernible until 5 days after induction. Osteopontin was observed 8 days after induction and found in association with calcified areas.The data revealed that inhibitors of crystallization like phytate are important in minimizing hydroxyapatite crystal formation in the first steps of soft tissue calcification. Histological results negate the involvement of osteopontin during initiation of soft tissue calcification. Osteopontin is probably involved in the control rather than the genesis of calcification (Grases et al., 2008a).

Inhibition of Development of Renal Calculi

A combination of nicotine-induced hypertension and vitaminD-induced hypercalcemia was used by Grases et al. (2007) to induce renal calcification in male rats fed a purified phytate-free diet. Rats without receiving phytate treatment exhibited calcium deposits in kidneys and papillae, and in renal tubules and vessels. Calcium deposits were not found in rats treated with phytate. Fragments of hydroxyapatite (HAP) calculi can induce the growth of calcium salts on their surfaces.The presence of phytate in synthetic urine at a concentration of 1.5 mg/L prevented calcium oxalate monohydrate formation on HAP renal calculi under normocalciuric conditions. Hence phytate inhibits crystallization both in the intrapapillary tissue and urine (Grases et al., 2007).

Reduction of Age-Related Cardiovascular Calcification

Grases et al. (2008b) demonstrated that dietary phytate treatment of rats from 10 weeks to 76 weeks of age significantly reduced age-related aorta calcification during aging.

Inhibition of Pericardium Calcification

The majority of the extracellular IP6 in mammalian organs, tissues and biological fluids originates from the diet and is not produced endogenously.On the other hand, intracellular IP6 probably originates in the cells. Little dietary IP6 absorption takes place during intestinal transit.Depletion of extracellular IP6 occurs rapidly when diets are poor in IP6. The data suggest that health benefits associated with extracellular IP6 are related to dietary IP6 (Grases et al., 2005).

Conclusion

IP6 is abundant in cereals and legumes. IP6 and lower inositol phosphates, in particular IP3, participate in important intracellular processes. IP6 manifests significant health benefits, such as anti-cancer effect, inhibition of formation of renal stones, and reduction of circulatory cholesterol (Grases et al., 2005). It is anticipated that some of the aforementioned health benefits can be acquired in a diet abundant in

IP6 and related compounds and these health benefits can in the future be exploited in the pharmacotherapy of diseases.

References

Anekonda TS, Wadsworth TL, Sabin R, Frahler K, Harris C, Petriko B, Ralle M, Woltjer R, Quinn JF. Phytic acid as a potential treatment for alzheimer's pathology: evidence from animal and in vitro models. *J Alzheimers Dis.* 2011;23(1):21-35.

Arriero Mdel M, Ramis JM, Perelló J, Monjo M. Inositol hexakisphosphate inhibits osteoclastogenesis on RAW 264.7 cells and human primary osteoclasts. *PLoS One*. 2012;7(8):e43187.

Colodny L, Hoffman RL. Inositol--clinical applications for exogenous use.*Altern Med Rev.* 1998;3(6):432-47.

Fux M, Benjamin J, Belmaker RH.Inositol versus placebo augmentation of serotonin reuptake inhibitors in the treatment of obsessive-compulsive disorder: a double-blind cross-over study. *Int J Neuropsychopharmacol.* 1999;2(3):193-195.

Grases F, Simonet BM, Vucenik I, Perelló J, Prieto RM, Shamsuddin AM. Effects of exogenous inositol hexakisphosphate (InsP(6)) on the levels of InsP(6) and of inositol trisphosphate (InsP(3)) in malignant cells, tissues and biological fluids. *Life Sci*. 2002;71(13):1535-46.

Grases F, Costa-Bauza A, Prieto RM. Intracellular and extracellular myo-inositol hexakisphosphate (InsP6), from rats to humans. *Anticancer Res.* 2005;25(3c):2593-7.

Vucenik I, Shamsuddin AM. Protection against cancer by dietary IP6 and inositol. *Nutr Cancer*. 2006;55(2):109-25.

Grases F, Isern B, Sanchis P, Perello J, Torres JJ, Costa-Bauza A. Phytate acts as an inhibitor in formation of renal calculi. *Front Biosci*. 2007;12:2580-7.

Grases F, Perelló J, Sanchis P, Isern B, Prieto RM, Costa-Bauzá A, Santiago C, Ferragut ML, Frontera G. Anticalculus effect of a triclosan mouthwash containing phytate: a double-blind, randomized, three-period crossover trial. *J Periodontal Res*. 2009;44(5):616-21.

Grases F, Prieto RM, Sanchis P, Saus C, De Francisco T. Role of phytate and osteopontin in the mechanism of soft tissue calcification. *J Nephrol.* 2008a;21(5):768-75

Grases F, Sanchis P, Perello J, Isern B, Prieto RM, Fernandez-Palomeque C, Saus C. Phytate reduces age-related cardiovascular calcification. *Front Biosci.* 2008b;13:7115-22.

Grases F, Sanchis P, Costa-Bauzá A, Bonnin O, Isern B, Perelló J, Prieto RM. Phytate inhibits bovine pericardium calcification in vitro. *Cardiovasc Pathol.* 2008c ;17(3):139-45.

Holub BJ.The nutritional significance, metabolism, and function of myo-inositol and phosphatidylinositol in health and disease. *Adv Nutr Res*. 1982;4:107-41.

Kapral M, Wawszczyk J, Hollek A, Weglarz L.Induction of the expression of genes encoding TGF-beta isoforms and their receptors by inositol hexaphosphate in human colon cancer cells. *Acta Pol Pharm.* 2013;70(2):357-63.

Khatiwada J, Verghese M, Davis S, Williams LL. Green tea, phytic acid, and inositol in combination reduced the incidence of azoxymethane-induced colon tumors in Fisher 344 male rats. *J Med Food.* 2011 ;14(11):1313-20.

Kunyanga CN, Imungi JK, Okoth MW, Biesalski HK, Vadivel V. Antioxidant and type 2 diabetes related functional properties of phytic acid extract from Kenyan local food ingredients: effects of traditional processing methods. *Ecol Food Nutr*. 2011;50(5):452-71

Larzelere MM, Campbell JS, Robertson M. Complementary and alternative medicine usage for behavioral health indications. *Prim Care.* 2010;37(2):213-36.

Levine J. Controlled trials of inositol in psychiatry. *Eur Neuropsychopharmacol.* 1997;7(2):147-55.

Liu N, Ru YJ, Cowieson AJ, Li FD, Cheng XCh. Effects of phytate and phytase on the performance and immune function of broilers fed nutritionally marginal diets. *Poult Sci*. 2008 Jun;87(6):1105-11.

Matsumoto H, Ito K, Yonekura K, Tsuda T, Ichiyanagi T, Hirayama M, Konishi T. Enhanced absorption of anthocyanins after oral administration of phytic acid in rats and humans. *J Agric Food Chem*. 2007;55(6):2489-96.

Nemets B, Fux M, Levine J, Belmaker RH.Combination of antidepressant drugs: the case of inositol. *Hum Psychopharmacol*. 2001;16(1):37-43.

Norhaizan ME, Ng SK, Norashareena MS, Abdah MA. Antioxidant and cytotoxicity effect of rice bran phytic acid as an anticancer agent on ovarian, breast and liver cancer cell lines. *Malays J Nutr*. 2011;17(3):367-75

Rizvi I, Riggs DR, Jackson BJ, Ng A, Cunningham C, McFadden DW. Inositol hexaphosphate (IP6) inhibits cellular proliferation in melanoma. *J Surg Res*. 2006;133(1):3-6.

Singh RP, Agarwal R.Prostate cancer and inositol hexaphosphate: efficacy and mechanisms. *Anticancer Res*. 2005;25(4):2891-903.

Wawszczyk J, Kapral M, Hollek A, Weglarz L.The effect of phytic acid on the expression of NF-kappaB, IL-6 and IL-8 in IL-1beta-stimulated human colonic epithelial cells. *Acta Pol Pharm*. 2012;69(6):1313-9.

Wawszczyk J, Orchel A, Kapral M, Hollek A, Weglarz L. Phytic acid down-regulates IL-8 secretion from colonic epithelial cells by influencing mitogen-activated protein kinase signaling pathway. *Acta Pol Pharm*. 2012;69(6):1276-82.

In: Inositol
Editors: H. Rocha and M. Cardoso
ISBN: 978-1-62808-520-4

Chapter 9

Cellular Functions of Inositol

Tzi Bun Ng and Charlene Cheuk Wing Ng[*]
School of Biomedical Sciences, Faculty of Medicine,
The Chinese University of Hong Kong, New Territories, Hong Kong,China

Abstract

Calcium-mobilizing receptors stimulate the hydrolysis of phosphatidylinositol 4,5-bisphosphate to 1,2-diacylglycerol and inositol 1,4,5-trisphosphate(Ins 1,4,5P3). Ins 1,4,5P3 released to the cytosol serves as a second messenger to release calcium from the endoplasmic reticulum. Growth factors such as platelet-derived growth factor, bombesin and vasopressin in fibroblasts or antigen in lymphocytes use the aforementioned transduction mechanism to produce intracellular mitogenic signals. The signaling mechanism via histamine H1-receptors is also used to mediate endothelial cell functions. Insulin regulates the phosphatidylinositol(PI) 3-kinase involving a physical association between the insulin receptor and the PI 3-kinase and tyrosine phosphorylation. Inositol is a nutrient required for keratinocyte proliferation. The Ins1,4,5P3/Ca2+ signal pathway functions to regulate such diverse processes as egg maturation and fertilization, growth, secretion, metabolism, neural activity, and excitation-contraction coupling in skeletal muscle. Myo-inositol serves as a clinically relevant osmolyte in the CNS, and its hexakisphosphate and pyrophosphorylated derivatives may play roles in such diverse cellular functions as DNA

[*] Email:bo21770@mailserv.cuhk.edu.hk, fax;852 26035123, phone:852 39436872.

repair, nuclear RNA export and synaptic membrane trafficking. Glycerophosphoinositols induce cell proliferation in thyroid cells, modulate actin cytoskeleton organization in fibroblasts, and reduce the invasive potential of tumour cell lines. Glycerophosphoinositols promote cytokine-dependent chemotaxis in T-lymphocytes induced by SDF-1alpha-receptor activation.

Keywords: Inositol, cellular functions, calcium

Introduction

This review article aims at reviewing the cellular functions of inositol which is present ubiquitously in the human body.The fundamental importance of inositol to cellular activities is discussed.

Regulation of Ion Channels

Membrane and across the membranes of intracellular organelles, especially the endoplasmic reticulum. The receptors stimulate the hydrolysis of an inositol lipid to produce a couple of second messengers. Diacylglycerol stays within the plasma membrane and activates protein kinase C resulting in phosphorylation of proteins which may control ionic channels, such as the calcium-dependent potassium channel or the Na+/H+ exchanger which regulates intracellular pH. The inositol trisphosphate (Ins 1,4,5P3) released to the cytosol serves as a second messenger and brings about calcium release from the endoplasmic reticulum. Ins 1,4,5P3 acts on a specific receptor to increase the passive efflux of calcium without any effect on the active calcium pump. This Ins 1,4,5P3-induced release of calcium from an internal membrane store may account for excitation-contraction coupling in skeletal muscle.Skinned skeletal muscle cells can be induced to contract by adding Ins 1,4,5P3. (Berridge, 1986a, Hughes and Putney, 1988).

The Ca2+ mobilizing actions of (1,4,5)IP3 are put to an end by two metabolic pathways, successive dephosphorylation to inositol bisphosphate (IP2), inositol monophosphate (IP) and inositol or by phosphorylation to inositol tetrakisphosphate (IP4) and successive dephosphorylation to different inositol phosphates. (Hughes and Putney, 1988).

Normal cells depend on growth factors for their multiplication. Growth factors such as platelet-derived growth factor, bombesin and vasopressin in fibroblasts or antigen in lymphocytes employ a specific inositol lipid to produce intracellular mitogenic signals They promote the hydrolysis of phosphatidylinositol 4,5-bisphosphate to yield diacylglycerol (DG) and inositol 1,4,5-trisphosphate (Ins1,4,5P3). DG activates protein kinase C, which raises intracellular pH by turning on a Na+/H+ exchanger. The other product, Ins1,4,5P3, serves as a second messenger which mobilizes calcium from intracellular stores. The rise in pH and calcium causes the onset of DNA synthesis (Berridge, 1986b).

Antioxidant Activity

The antioxidant property of inositol is the consequence of its quenching action on reactive oxygen, intermediates and formation of conjugates with compounds like arachidonic acid and the antiglycating property owing to scavenging of glucose. The antioxidant and the antiglycating properties of inositol may delay or prevent cataract formation (Ramakrishnan et al., 1999).

Iron chelates of inositol 1,2,3-trisphosphate and inositol 1,2,3,6-tetrakisphosphate are devoid of coordination sites and inhibit iron-catalyzed oxidation of ascorbic acid and peroxidation of arachidonic acid.In contrast, iron chelates of inositol 1,2,6-trisphosphate and inositol 1,2,5,6-tetrakisphosphate possess coordination sites, allow iron-catalyzed ascorbic acid oxidation, and enhance arachidonic acid peroxidation. The 1,2,3-trisphosphate grouping of inositol hexakisphosphate inhibit iron-catalyzed hydroxyl radical formation.Hence, the antioxidant potential of inositol 1,2,3-trisphosphate and inositol 1,2,3,6-tetrakisphosphate in cells and other biological systems may be reinforced by the resistance of their iron chelates to enzymatic hydrolysis of the functional 1,2,3-trisphosphate array (Phillippy and Graf, 1997).

Cellular Functions

Inositol pyrophosphates manifest an array of cellular functions, including vesicular trafficking, apoptosis, telomere maintenance and protein phosphorylation. The activity of inositol pyrophosphates appears to be related to their rapid turnover in cells and also to their pyrophosphate groups, which have high energy bonds. The observations suggest that inositol pyrophosphates are a class of cellular messengers (Bennett et al., 20060.

One function of novel phosphoinositides and their metabolites may reside at the level of cytoskeletal and cell shape regulation (Wasilenko, 19920.

The inositol trisphosphate (Ins1,4,5P3)/Ca2+ signalling system has been adapted to regulate a diversity of cellular processes comprising egg maturation and fertilization, development, growth, secretion, metabolism, neural activity, and perhaps excitation-contraction coupling in skeletal muscle, proliferation and apoptosis (Berridge, 1986, 1997).

Cellular Metabolism and Biological Functions

Glycerophosphoinositols manifest a plethora of actions in a variety of cell types comprising induction of thyroid cell proliferation, modulation of actin cytoskeletal organization in fibroblasts, and diminution of the invasive potential of tumor cell lines. More recent investigations include their effects in inflammatory and immune responses. Glycerophosphoinositols display cytokine-dependent chemotaxis in T-lymphocytes induced by SDF-1alpha-receptor activation, signifying roles played by these compounds as modulators of T-cell signalling and T-cell responses (Corda et al., 2009).

Glycerophosphoinositols regulate cell proliferation in the thyroid and actin cytoskeleton organization in fibroblasts and lymphocytes. in immune and inflammatory responses (Corda et al., 2012).

Nuclear and Cytoplasmic Functions

Nuclear and cytoplasmic functions have been ascribed to inositol pyrophosphates, from intracellular trafficking to control of telomere length and from apoptosis regulation to augmentation of insulin secretion (Saiardi, 2012).

Co-Factors for Nuclear Processes

Phosphoinositides and their soluble analogues, inositol polyphosphates, have been shown to serve as essential co-factors for DNA repair, transcription regulation and RNA dynamics (Hammond et al., 2004).

Control of NK Cell Effector Functions

Natural killer (NK) cells have important functions in cancer immunosurveillance, BM allograft rejection, fighting infections, tissue homeostasis, and reproduction. NK cell-based therapies are promising treatments for blood cancers.inositol (1,3,4,5)tetrakisphosphate (IP(4)) IP(4) promotes NK cell terminal differentiation and acquisition of a mature NKR repertoire.Nevertheless,in mature NK cells, IP(4) limits NKR-induced IFNγ secretion, granule exocytosis, and target-cell killing, in part by inhibiting the PIP(3) effector-kinase Akt. This identifies IP(4) as an important novel regulator of NK cell development and function and expands our understanding of the therapeutically important mechanisms dampening NK cell responses (Sauer et al., 2013).

Control of Reproductive Development

Carlomagno et al. (2011) summarized the functions of myo-inositol in oocyte quality and development, sperm quality, maintenance of pregnancy and embryo development

Control of Glutamatergic Transmission

In the adult brain, maintenance of synaptic functions is crucial for processing of neuronal information. Astrocytes express glutamate transporters that expediently remove glutamate from the extracellular space and play a critical role in glutamatergic transmission. The inositol 1,4,5-trisphosphate signaling cascade regulates the glutamate/aspartate transporter levels in Bergmann glial cells to maintain glutamate clearance in the cerebellum (Mashimo et al., 2010).

Functions in the Central Nervous System (CNS)

Inositol phospholipids and inositol phosphates participate in signal transduction and Ca2+ homeostasis in the CNS as well as non-nervous tissues. Myoinositol is a clinically relevant osmolyte in the CNS, and its hexakisphosphate and pyrophosphorylated derivatives are probably involved in DNA repair, nuclear RNA export and synaptic membrane trafficking (Fisher et al., 2002).

Costa (1994) reviewed some of the concepts of receptor—activated phosphoinositide metabolism during CNS development. The acetylcholine muscarinic receptor—stimulated inositol metabolism is a potential target for the developmental neurotoxicity of ethanol.

Signal Transduction in Clonal Pituitary Cells

Clonal rat pituitary tumor cells respond to thyrotropin-releasing hormone (TRH) by interaction of the hormone-occupied TRH receptor on the cells with a GTP-binding protein. This is then followed by phosphodiesterase-catalyzed hydrolysis of phosphatidylinositol 4,5-bisphosphate to the intracellular messengers inositol 1,4,5-trisphosphate and 1,2-diacylglycerol. The ensuing alterations in cytosolic free calcium follow a complex temporal pattern. These changes, together with activation of protein kinase C, can regulate prolactin secretion from the pituitary cells (Drummond, 1986).

Nutrient for Keratinocyte Growth

Bovine hypothalamic myo-inositol promotes growth in cultured human epidermal keratinocytes. No other cutaneously derived cell type examined was affected by inositol (Gordon et al., 1988).

Inositol Metabolism in Endothelial Cells

Stimulation of hydrolysis of inositol lipid is the transmembrane signaling mechanism utilized by histamine H1-receptors in mediating various endothelial cell functions (Lo and Fan, 1987).

Inositol Metabolism in Platelets

In human platelets, inositol (1,4,5) triphosphate metabolism is regulated by Ca2+-calmodulin, and its metabolites may play a role as second messengers (Daniel et al., 1988).

Insulin, Diabetes and Inositol Metabolism

In diabetes mellitus, enhanced inositol phosphate accumulation may bring about accelerated platelet functions (Ishii et al., 1991).Changes in myoinositol metabolism, polyol content, and Na+/K+ pump activity in cultured neuroblastoma cells are induced by hyperglycemia and restoration to the normal is observed following return to normoglycemia (Yorek et al., 1993).

Insulin action appears to require the protein-tyrosine kinase domain of the beta subunit of the insulin receptor. insulin at physiological concentrations regulates the PI 3-kinase and suggest that this regulation involves a physical association between the insulin receptor and the PI 3-kinase and tyrosyl phosphorylation Ruderman et al., 1990.

Inositol phosphate -oligosaccharides have been suggested as the second messengers of certain effects of insulin because they mimic the metabolic effects of insulin in fat cells. A role of mannose- and glucosamine-containing inositol phosphate-oligosaccharides as second messengers in insulin-stimulated lipogenesis is suggested by the observation that inositol monophosphate, mannose, and glucosamine antagonize this effect of insulin (Machicao et al., 1990).

Effect of Inositol Depletion

The inositol-depletion hypothesis was proposed to account for the mechanism of therapeutic action of mood-stabilizing drugs. Inositol and inositol-containing compounds, encompassing phosphoinositides and inositol phosphates, exhibit signalling roles and regulatory functions in diverse cellular processes. This suggests that depleting inositol may disrupt a great variety of cellular functions. Some of the affected functions may be related to bipolar disorders (Deranieh and Greenberg 2009.

Mice Lacking in Inositol Hexakisphosphate Kinase 1

Changes in phenotype in mice deficient in inositol hexakisphosphate kinase 1 (IP_6K1) uncover signaling functions for these molecules in immunological functions, insulin homeostasis, and obesity. Inositol pyrophosphates regulate these processes by suppressing activation of the serine-threonine kinase Akt. Parallel investigations of IP_6K2 establish this enzyme as a cell death inducer acting by stimulating the proapoptotic protein p53. Inositol polyphosphate multikinase (IPMK) catalyzes the formation of inositol phosphate IP_5 but also possesses phosphatidylinositol 3-kinase activity which participates in activation of Akt (Chakraborty et al., 2011).

Conclusion

The foregoing review disclosed that inositol displays a repertoire of cellular functions which range from signal transduction pathways of platelet-derived growth factor, bombesin and vasopressin, cytoskeletal and cell shape regulation,reproductive development, neurotransmission, vesicular trafficking, apoptosis, telomere maintenance and protein phosphorylation. A diversity of cell types including keratinocytes, endothelial cells, platelets, and pituitary cells have active inositol metabolism.Continued research will unveil more functions of inositol which may be exploited to the welfare of mankind.

References

Berridge MJ. Regulation of ion channels by inositol trisphosphate and diacylglycerol. *J Exp Biol.* 1986a;124:323-35.

Berridge MJ. Growth factors, oncogenes and inositol lipids. *Cancer Surv.* 1986b;5(2):413-30.

Berridge MJ. Inositol trisphosphate and calcium mobilization. *Ciba Found Symp.* 1986;122:39-57.

Berridge MJ. The 1996 Massry Prize. Inositol trisphosphate and calcium: two interacting second messengers. *Am J Nephrol.* 1997;17(1):1-11.

Carlomagno G, Nordio M, Chiu TT, Unfer V. Contribution of myo-inositol and melatonin to human reproduction. *Eur J Obstet Gynecol Reprod Biol.* 2011;159(2):267-72.

Chakraborty A, Kim S, Snyder SH. Inositol pyrophosphates as mammalian cell signals.*Sci Signal.* 2011;4(188):re1.

Corda D, Zizza P, Varone A, Bruzik KS, Mariggiò S. The glycerophosphoinositols and their cellular functions. *Biochem Soc Trans.* 2012;40(1):101-7.

Bennett M, Onnebo SM, Azevedo C, Saiardi A. Inositol pyrophosphates: metabolism and signaling. *Cell Mol Life Sci.* 2006;63(5):552-64.

Corda D, Zizza P, Varone A, Filippi BM, Mariggiò S. The glycerophosphoinositols: cellular metabolism and biological functions. *Cell Mol Life Sci.* 2009;66(21):3449-67.

Costa LG.Signal transduction mechanisms in developmental neurotoxicity: the phosphoinositide pathway. *Neurotoxicology.* 1994;15(1):19-27.

Daniel JL, Dangelmaier CA, Smith JB. Calcium modulates the generation of inositol 1,3,4-trisphosphate in human platelets by the activation of inositol 1,4,5-trisphosphate 3-kinase. *Biochem J.* 1988;253(3):789-94.

Deranieh RM, Greenberg ML. Cellular consequences of inositol depletion. *Biochem Soc Trans.* 2009;37(Pt 5):1099-103.

Drummond AH. Inositol lipid metabolism and signal transduction in clonal pituitary cells. *J Exp Biol.* 1986;124:337-58.

Fisher SK, Novak JE, Agranoff BW. Inositol and higher inositol phosphates in neural tissues: homeostasis,metabolism and functional significance. *J Neurochem.* 2002;82(4):736-54.

Gordon PR, Mawhinney TP, Gilchrest BA. Inositol is a required nutrient for keratinocyte growth. *J Cell Physiol.* 1988;135(3):416-24.

Hammond G, Thomas CL, Schiavo G. Nuclear phosphoinositides and their functions. *Curr Top Microbiol Immunol.* 2004;282:177-206.

Hughes AR, Putney JW Jr. Metabolism and functions of inositol phosphates. *Biofactors.* 1988;1(2):117-21.

Ishii H, Umeda F, Hashimoto T, Nawata H. Increased inositol phosphate accumulation in platelets from patients with NIDDM. *Diabetes Res Clin Pract.* 1991;14(1):21-7.

Machicao F, Mushack J, Seffer E, Ermel B, Häring HU. Mannose, glucosamine and inositol monophosphate inhibit the effects of insulin on lipogenesis. Further evidence for a role for inositol phosphate-oligosaccharides in insulin action. *Biochem J.* 1990;266(3):909-16.

Mashimo M, Okubo Y, Yamazawa T, Yamasaki M, Watanabe M, Murayama T, Iino M.Inositol 1,4,5-trisphosphate signaling maintains the activity of glutamate uptake in Bergmann glia. *Eur J Neurosci.* 2010;32(10):1668-77.

Lo WW, Fan TP. Histamine stimulates inositol phosphate accumulation via the H1-receptor in cultured human endothelial cells. *Biochem Biophys Res Commun.* 1987;148(1):47-53.

Phillippy BQ, Graf E. Antioxidant functions of inositol 1,2,3-trisphosphate and inositol 1,2,3,6-tetrakisphosphate. *Free Radic Biol Med.* 1997;22(6):939-46.

Ramakrishnan S, Sulochana KN, Punitham R. Two new functions of inositol in the eye lens: antioxidation and antiglycation and possible mechanisms. *Indian J Biochem Biophys.* 1999;36:129-33.

Ruderman NB, Kapeller R, White MF, Cantley LC. Activation of phosphatidylinositol 3-kinase by insulin. *Proc Natl Acad Sci U S A.* 1990;87(4):1411-5.

Saiardi A. Cell signalling by inositol pyrophosphates. *Subcell Biochem.* 2012;59:413-43.

Sauer K, Park E, Siegemund S, French AR, Wahle JA, Sternberg L, Rigaud S, Jonsson AH, Yokoyama WM, Huang YH.Inositol tetrakisphosphate limits NK cell effector functions by controlling PI3K signaling. *Blood.* 2013;121(2):286-97.

Song F, Zhang J, Zhao Y, Chen W, Li L, Xi Z. Synthesis and antitumor activity of inositol phosphotriester analogues. *Org Biomol Chem.* 2012;10(18):3642-54.

Szijgyarto Z, Garedew A, Azevedo C, Saiardi A. Influence of inositol pyrophosphates on cellular energy dynamics. *Science.* 2011;334(6057):802-5.

Wasilenko WJ. Phosphoinositides and cell growth. *Adv Exp Med Biol.* 1992;321:147-51.

Yorek MA, Dunlap JA, Stefani MR, Davidson EP. Reversal of hyperglycemic-induced defects in myo-inositol metabolism and Na+/K+ pump activity in cultured neuroblastoma cells by normalizing glucose levels. *Metabolism.* 1993;42(9):1180-9.

Index

#

A

B

C

D

E

F

G

H

I

J

K

L

M

N

O

P

Q

R

S

U

V

W

X

Y

Z